T0313743

# Innovative Processing and Synthesis of Ceramics, Glasses, and Composites V

## Related titles published by The American Ceramic Society:

*Boing-Boing the Bionic Cat and the Jewel Thief*
By Larry L. Hench
©2001, ISBN 1-57498-129-3

*Innovative Processing and Synthesis of Ceramics, Glasses, and Composites IV*
*(Ceramic Transactions, Volume 115)*
Edited by Narottam P. Bansal and J.P. Singh
©2000, ISBN 1-57498-111-0

*Innovative Processing and Synthesis of Ceramics, Glasses, and Composites III*
*(Ceramic Transactions Volume 108)*
Edited by J.P. Singh, Narottam P. Bansal, and Koichi Niihara
©2000, ISBN 1-57498-095-5

*Advances in Ceramic Matrix Composites V (Ceramic Transactions, Volume 103)*
Edited by Narottam P. Bansal, J.P. Singh, and Ersan Ustundag
©2000, ISBN 1-57498-089-0

*The Magic of Ceramics*
By David W. Richerson
©2000, ISBN 1-57498-050-5

*Boing-Boing the Bionic Cat*
By Larry L. Hench
©2000, ISBN 1-57498-109-9

*Ceramic Innovations in the 20th Century*
Edited by John B. Wachtman Jr.
©1999, ISBN 1-57498-093-9

*Advances in Ceramic Matrix Composites IV (Ceramic Transactions, Volume 96)*
Edited by J.P. Singh and Narottam P. Bansal
©1999, 1-57498-059-9

*Innovative Processing and Synthesis of Ceramics, Glasses, and Composites II*
*(Ceramic Transactions, Volume 94)*
Edited by Narottam P. Bansal and J.P. Singh
©1999, ISBN 1-57498-060-2

*Innovative Processing and Synthesis of Ceramics, Glasses, and Composites*
*(Ceramic Transactions, Volume 85)*
Edited by Narottam P. Bansal, Kathryn V. Logan, and J.P. Singh
©1998, ISBN 1-57498-030-0

*Advances in Ceramic Matrix Composites III (Ceramic Transactions, Volume 74)*
Edited by Narottam P. Bansal and J.P. Singh
©1996, ISBN 1-57498-020-3

For information on ordering titles published by The American Ceramic Society, or to request a publications catalog, please contact our Customer Service Department at 614-794-5890 (phone), 614-794-5892 (fax), <customersrvc@acers.org> (e-mail), or write to Customer Service Department, 735 Ceramic Place, Westerville, OH 43081, USA.

Visit our on-line book catalog at <www.ceramics.org>.

Ceramic Transactions
Volume 129

# Innovative Processing and Synthesis of Ceramics, Glasses, and Composites V

*Proceedings of the Innovative Processing and Synthesis of Ceramics, Glasses, and Composites symposium held at the 103rd Annual Meeting of The American Ceramic Society, April 22–25, 2001, in Indianapolis, Indiana, USA.*

*Edited by*

**J.P. Singh**
Argonne National Laboratory

**Narottam P. Bansal**
National Aeronautics and Space
Administration Glenn Research Center

**Amit Bandyopadhyay**
Washington State University

*Published by*

**The American Ceramic Society**
735 Ceramic Place
Westerville, Ohio 43081
www.ceramics.org

*Proceedings of the Innovative Processing and Synthesis of Ceramics, Glasses, and Composites symposium held at the 103rd Annual Meeting of The American Ceramic Society, April 22–25, 2001, in Indianapolis, Indiana, USA.*

*Cover photo: "The SEM images show typical morphology of the spray particles used in this work," is courtesy of R. Gadow, A. Killinger, and C. Li, and appears as figure 4a in the paper "Plasma Sprayed Ceramic Coatings on Glass and Glass Ceramic Substrates," which begins on page 15.*

Library of Congress Cataloging-in-Publication Data

A CIP record for this book is available from the Library of Congress.

For information on ordering titles published by The American Ceramic Society, or to request a publications catalog, please call 614-794-5890.

Printed in the United States of America.

4  3  2  1–05  04  03  02

ISSN 1042-1122
ISBN 1-57498-137-4

# Contents

# Reaction Forming

# Functionally Graded Materials & Coatings

# Laminated Object Manufacturing

# Electronic and Magnetic Materials

# Preface

This volume contains papers presented at a symposium on Innovative Processing and Synthesis of Ceramics, Glasses and Composites held during the 103rd Annual Meeting and Exposition of the American Ceramic Society in Indianapolis, April 22-25, 2001. This symposium provided an international forum for scientists and engineers to discuss all aspects of processing and synthesis of ceramics, glasses and composites. A total of 114 papers, including invited talks, oral presentations, and posters, were presented from 17 countries (the United states, Belgium, Brazil, Canada, Germany, Iran, Israel, Italy, Japan, Mexico, the People's Republic of China, Republic of Korea, Spain, Switzerland, Taiwan, United Kingdom, and Venezuela). The speakers represented universities, industry, and research laboratories.

This volume contains 18 invited and contributed papers, all peer-reviewed according to American Ceramic Society procedures. The latest developments in processing and characterization are covered: CVD/CVI and plasma technology, gas infiltration and polymer processing, rheological behavior, mechanical alloying, reaction forming, functionally graded materials and coatings, laminated object manufacturing, and electronic and magnetic materials. All of the most important aspects necessary for understanding and further development of ceramic processing and characterization are discussed.

The organizers are grateful to all participants and session chairs for their time and effort, to authors for their timely submissions and revisions of the manuscripts, and to reviewers for their valuable comments and suggestions; without the contributions of all involved, this volume would not have been possible. Financial support from the American Ceramic Society is gratefully acknowledged. Thanks are due to the staff of the Meetings and Publications Department of the American Ceramic Society for their tireless efforts. Especially, we greatly appreciate the helpful assistance and cooperation of Sarah Godby throughout the production process of this volume.

We hope that this volume will serve as a useful reference for professionals working in the field of synthesis and processing of ceramics, glasses, and composites.

J. P. Singh
Narottam P. Bansal
Amit Bandyopadhyay

# CVD/CVI and Plasma Technology

# CUBIC AND TETRAGONAL HAFNIUM OXIDE FILMS PREPARED BY ION BEAM ASSISTED DEPOSITION

Rafael R. Manory, Ippei Shimizu,
Takanori Mori and Shoji Miyake
Joining and Welding Research Institute
Osaka University, 11-1 Mihagaoka,
Ibaraki, Osaka 547-0067 Japan

Hidenori Saito
Kanagawa High-Technology
Foundation (KTF), 3-2-1 Sakado
Takatsu-ku, Kawasaki- shi
Kanagawa 213-0012 Japan

Giora Kimmel
Dept. of Materials Engineering,
Ben Gurion University of the Negev
Beer-Sheva, 84307 Israel

Takeo Tanaka
Dept. of Mechanical Engineering
Osaka Sangyo University
3-1-1 Nakagaito, Daito, Osaka,
574, Japan

## ABSTRACT
HfO$_2$ films were prepared by ion beam assisted deposition (IBAD) of hafnium with simultaneous bombardment with oxygen ions, accelerated between 1-20keV. The ratio between the hafnium arrival rate and the oxygen ion dose, (transport ratio, TR) was varied between 0.5 and 10. Structure and composition were characterized using XRD, XPS, RBS, EPMA.

Two unexpected structures were observed with variation in parameters, a cubic CaF$_2$-type structure with a parameter of about 0.412nm and a new tetragonal structure, believed to be different from the high temperature HfO$_2$ structure. The composition of the films was found to be under-stoichiometric, with Hf/O ratio around 1.5. The new tetragonal structure appears to be correlated to low oxygen content, below 1.5. Film orientation could be controlled by varying the substrate rotation during deposition.

## INTRODUCTION
Hafnium dioxide has a number of isomorphs and has been the topic of numerous studies. At room temperature the equilibrium phase of hafnia is monoclinic baddeleyte,[1-3] which has the lowest free energy of formation and the largest volume. The effects of stoichiometry and temperature variations under atmospheric and higher pressures have been studied by various authors, and the phase diagram of Hf-O has been determined.[1-3] There is some solubility of O in Hf, but the only oxide phase identified in the phase diagram at room temperature

was $HfO_2$, with a phase boundary at about 63.5 at% oxygen.[3] At about 1300K, monoclinic $HfO_2$ transforms into a tetragonal structure and at about 2700K into a cubic structure,[3,4] of the $CaF_2$-type[4]. It is interesting to note that Passerini[5] identified the $CaF_2$ structure in the $HfO_2$-$CeO_2$ system as early as 1930. With increasing pressure, densification of monoclinic $HfO_2$ occurs by transformation into a number of orthorhombic phases, which have been closely studied and mapped.[1,2,6,7].

With the exception of the main phases obtained at atmospheric pressure, other phases were observed in bulk material only under high-pressure conditions. The high-density phases obtained under high-pressure conditions appear to have elastic moduli of the order of 145GPa and are therefore considered among the super-hard materials[8].

The main interest however in this material is for optical coatings. It presents a relatively high laser damage threshold due to its high melting point, thermal and chemical stability [9,10] and a large transparent range from the IR to the UV (0.22-12μm), with a relatively high refractive index,[10] and is also relatively easy to obtain by evaporation. In view of these properties, a number of techniques were used to produce such films, including sol-gel[11], atomic layer deposition (ALD)[12], in addition to vacuum deposition methods.

Among the various attempts to obtain high quality optical films, ion beam assisted deposition (IBAD) has also been used in a number of studies, [9-13] with various ion sources, sometimes using ionic species other than oxygen. In connection to the present work, is worth noting that non-stoichiometry of the $HfO_x$ films has been reported[14], but without effects on the structure.

The cubic and tetragonal high temperature structures have been sometimes observed in thin films in small amounts alongside the monoclinic structure. Aarik et al.[12] have reported the presence of the orthorhombic phase in films obtained by the ALD method at 500°C, whereas Ritala et al.[15,16] have observed weak peaks of a tetragonal structure. Recently, Aarik et al.[17] have reported the presence of cubic nano-crystallites on the surface layers of monoclinic $HfO_2$ films grown by the ALD method at temperatures in the vicinity of 900°C.

The work presented here was undertaken in an attempt to explore the effects of ion bombardment on phase transitions and properties of thin $HfO_2$ films deposited by IBAD at ion energies of 1-20 keV, significantly higher than ion acceleration energies used in other[9,13,14] IBAD works. The bombarding ion was oxygen, but we varied the transport ratio (TR), defined as the ratio between the oxygen flow-rate, and the hafnium evaporation rate. The parameters varied included transport ratio, ion energy, alloying with Ce, substrate cooling and substrate rotation speed. In contradiction with other reports, we obtained films consisting not of a monoclinic phase with some other high temperature phase, but entirely, or almost entirely of either cubic or tetragonal phase, (which appears to belong to a different space group than the high temperature tetragonal structure of hafnia).

TABLE I – List of samples and experimental conditions

| Sample | Ion Energy | T.R. | Deposition rate | Rotation rate | Composition (O/Hf ratio) | |
|--------|-----------|------|-----------------|---------------|--------------------------|---|
| | keV | --- | --- | sec/pass | XPS | EPMA |
| 1 | 20 | 5 | 3.5 | low | 1.30 | 1.44 |
| 2 | 20 | 1 | 0.7 | low | 1.55 | 1.54 |
| 3 | 20 | 10 | 5.3 | low | 1.47 | 1.44 |
| 4 | 20 | 0.5 | 0.3 | low | 1.71 | --- |
| 6 | 20 | 4 | 2.8 | low | --- | --- |
| 7 | 20 | 2 | 1.3 | low | --- | --- |
| 8 | 1.0 | 5 | 1.8 | low | 1.74 | --- |
| 9 | 10 | 5 | 2.9 | low | 1.52 | --- |
| 11 | 5 | 5 | 2.7 | low | 1.72 | --- |
| 20* | 20 | 5 | 3.5 | low | --- | 1.56 |
| 14 | 20 | 5 | Hf = 3.8, Ce = 0.3 | low | --- | Ce/(Hf+Ce)=0.025 |
| 15 | 1 | 5 | Hf = 1.7, Ce = 0.2 | low | --- | Ce/(Hf+Ce)=0.021 |
| 16 | 10 | 5 | Hf = 3.9, Ce = 0.5 | low | --- | --- |
| 37 | 20 | 5 | 3.4 | 87 | --- | --- |
| 36 | 20 | 5 | 3.4 | 50 | --- | --- |
| 28 | 20 | 5 | 3.3 | 20 | --- | 1.27 |
| 35 | 20 | 5 | 3.3 | 0 | --- | 1.16 |

*This sample was deposited without water-cooling the substrate.

## EXPERIMENTAL

A compact IBAD system with a bucket-type 2.45-GHz electron-cyclotron-resonance ion source and an electron beam evaporation source was used for the preparation of the $HfO_2$ films. The basic details of the ion source have been described elsewhere.[18,19] The acceleration voltage of the extracted oxygen ions was varied in the range of 1-20 kV. The ion current density measured with a Faraday cup was typically 26-51 $\mu$A/cm$^2$. The base pressure, evacuated with a 1500-l/s cryo-pump was 1-2x $10^{-6}$ torr. The working pressure of oxygen during the ion source operation was 1-3x$10^{-4}$ torr due to gas flow from the ion source. The substrate was maintained at a low temperature by water-cooling the plate under the substrate holder. The films were obtained on Si (100) wafers by depositing Hf vapor (from a target with purity 98 at%, with the major impurity being Zr) under simultaneous bombardment with oxygen ions. The incidence angle of the oxygen ion beam was normal to the substrate and the hafnium vapor reached the substrate at an angle of approximately $45°$ from normal. The transport-rate ratio of was varied in the range of 0.5-10. Typical film thickness was in the range of 400-600nm, as measured by surface profilometry. The substrate was rotated during deposition at various speeds varying between 0-3 rpm. (One rotation in 20 sec. is 'high rate' and in 90 sec is considered 'slow').

X-ray diffraction (XRD) data was collected using Cu radiation in a Rigaku Miniflex apparatus at 30 kV-15 mA, with a Ni filter, in the $\theta/2\theta$ mode. The chemical composition was measured using electron probe microanalysis (EPMA). Selected samples were analyzed by X-ray photoelectron spectroscopy (XPS) and Rutherford backscattering (RBS). For standard $HfO_2$, a powder made by Alfa-Johnson Matthey Gmbh. (Puratronic - 99.988% purity), was used.
Microstructures were obtained (courtesy of Kanagawa High-Technology Foundation (KTF)) using a field emission scanning electron microscope (FE-SEM) capable of ultra high magnification (up to $x10^5$).

## RESULTS AND DISCUSSION

The experimental conditions for the films discussed here are given in Table I (organized in order of parameter variation). Figure 1a presents the XRD spectrum of sample #1 (20 keV, TR=5). Three strong peaks at about 30, 35 and $50°$ are observed, with additional small peaks at about 40 and $60°$. The three strong peaks together do not belong to any of the ICDD cards for $HfO_2$ structures. This spectrum, with the exception of the peak at $40°$ appeared to be typical of a cubic structure, fitting the $CaF_2$ structure reported by Passerini[5], and also mentioned in the Pearson handbook.[20] A calculated spectrum for such a structure with $a_o$=0.506nm is shown in Fig. 1b.

The peaks at $40°$ and $55°$ in Fig. 1a are broader than those at $30°$ and $35°$ (full width at half maximum -FWHM- is 0.52, compared to 0.32 for the latter two). The broad peaks were de-convoluted using the Gaussian formula and it was determined that this film has a tetragonal rather than cubic structure, with the parameters a=0.5055 and c=0.5111 nm (c/a ratio 1.011). The calculated spectrum of this structure is presented in Fig.1c. As can be observed, this spectrum fits well the spectrum shown in Fig. 1a. The c/a ratio in this structure differs from the value of 1.021 of the normal tetragonal hafnia structure of high-temperature (ICDD card 08-0342). The Miller indices listed in the ICDD card do not correspond to those observed in our film, and this fact, together with the change in the c/a ratio this leads us to believe that the structure presented here is different from the tetragonal high temperature $HfO_2$ phase.

On the other hand, by varying ion energy we could obtain films exhibiting the cubic structure. The XRD spectrum of sample no. 8 (1keV, TR5) is shown in Fig. 1d. The additional peak at $40°$ does not appear here and this spectrum is similar to the cubic XRD profile of Fig. 1b.

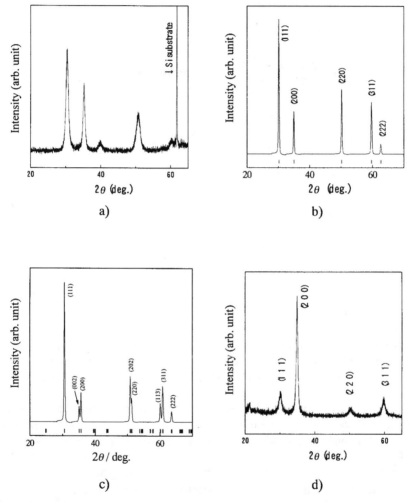

Figure 1. XRD spectra. a) Sample #1; b) Calculated spectrum of the $CaF_2$ structure with $a_o$=0.506nm; c) calculated spectrum of a tetragonal structure fitting Fig. 1a; d) Spectrum of sample #8, fitting the calculated cubic model of Fig. 1b.

A film with a monoclinic structure is shown in Fig. 2, where the XRD spectrum of sample no. 7 is compared to a standard sample for monoclinic $HfO_2$. The film also presents a certain degree of preferred orientation.

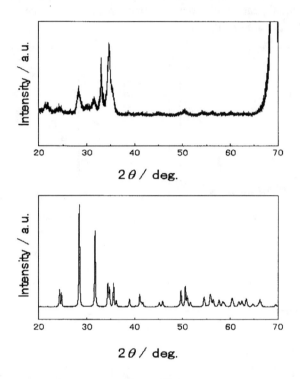

Fig. 2. XRD spectrum of sample #7 (20keV, T.R.2) (top), and comparison with the spectrum of standard monoclinic $HfO_2$ powder (bottom). The deposited film also contains a second phase –(either cubic or tetragonal, oriented (200)).

Figure 3 presents high-resolution FE-SEM micrographs of three types of films obtained with variation of parameters- samples nos.1, 7 and 14, which are almost entirely tetragonal monoclinic and cubic (oriented), respectively. It should be noted that a small component of a second phase is always present in these samples. All three samples present a columnar structure, but the grains and their spacing are very different. The mechanical properties of these samples are discussed elsewhere[21]. Sample no. 1 (Fig 3 -top) presents a very dense columnar structure and a smooth surface, whereas the monoclinic sample no. 7 (Fig.3-middle) shows

Innovative Processing and Synthesis of Ceramics

smaller and more irregular grains, with larger pores and a much rougher surface. The third sample, no. 14 (Fig. 3 bottom) is cubic, oriented (100), and was obtained by addition of a small amount of Ce, which was added to stabilize the cubic structure (based on the work of Passerini[5]).

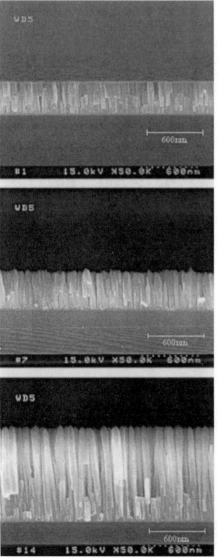

Fig. 3 FE-SEM micrographs of three types of films with different structures: top: sample no. 1- tetragonal; middle: sample no. 7 –monoclinic; bottom- sample no. 14 - cubic (oriented (200)).

The columnar structure is less dense, and the surface is less smooth than that of the tetragonal sample.

The composition of the films is given in the last column of Table I. All the films for which data is available are oxygen deficient. Given that the transport ratio was larger than 0.5 in most films, the oxygen deficiency is not surprising by itself. (The film grown at T.R.=0.5, 20keV was very thin and nearly amorphous and will not be discussed here). Others have already reported under-stoichiometric films obtained by IBAD[14]. However, the appearance of the tetragonal and cubic structures reported here is unusual and seems not to be correlated with composition. For example, sample no. 20 was grown at otherwise the same conditions as sample no. 1 (Fig. 1a), but without substrate cooling. The structure obtained in this case contains a large monoclinic component, as shown in Fig. 4. This leads to the conclusion that the formation of these unusual structures is associated with the fast cooling rate.

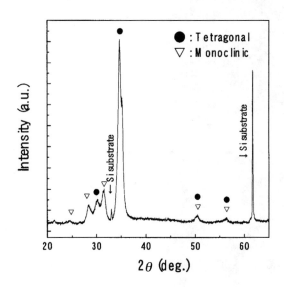

Fig. 4 XRD spectrum of sample no. 20 showing a large monoclinic component alongside the tetragonal phase. This sample was deposited without substrate cooling (see Table I ).

It is interesting to note that despite the well-known affinity of Hf to oxygen, these structures are stable in air at room temperature. In samples analyzed by XPS the oxygen content in the surface layer was found to correlate with EPMA results, and was lower in the subsurface layers (after sputtering). The films can be therefore described as having the formula $HfO_{2-x}$. One of the reasons for the oxygen deficiency at 20 keV is the fact that at this energy, part of the oxygen ions are implanted into the substrate and do not participate in the formation of hafnia. This was shown in RBS results of a sample deposited under the same conditions as sample no. 1, but on a carbon substrate (data not shown).

The orientation of the films is an important parameter in thin films deposition. The orientation of cubic and tetragonal films was found to be strongly affected by the sample rotation speed. Figure 5 presents spectra of films deposited at the same energy (20 keV) and at the same TR value the only variable being the substrate rotation speed. The intensity of the (200) peak varies among these films. In order to assess the preferred orientation the intensity of the (200) peak was measured against that of the (111) peak and the results are shown in Figure 6. It should be noted that in these series both cubic and tetragonal films were included. (We also obtained films, such as sample no. 14 for example, which have only the (200) peak in the XRD spectrum. Films deposited at a fast rotation speed are more oriented.

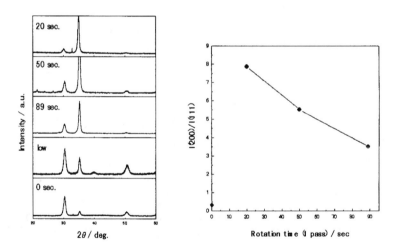

Figure 5 (left): The effect of rotation on film orientation for films deposited at 20 keV, TR=5 (without Ce addition).

Figure 6 (right): Variation of the intensity ratio I(200)/I (111) (indicating degree of preferred orientation) with rotation speed. The (200) orientation is enhanced with faster rotation.

## CONCLUSIONS

In this work we performed ion beam assisted deposition of hafnium oxide at ion energies of 1-20 keV and varying transport ratios. The main conclusions of this study are:

- The deposited films showed a columnar structure, with tetragonal, cubic, monoclinic or mixed structures observed with varying deposition parameters.
- A new tetragonal structure and a rarely observed cubic structure were repeatedly obtained.
- Substrate cooling was found to influence the formation of the tetragonal structure and to promote monoclinic growth.
- The films were oxygen deficient, $HfO_{2-x}$, with x varying between 0.3-0.8 approximately.
- Sample rotation was found to affect film orientation in cubic and tetragonal films, and can be used for tailoring the structure.

Work is still in progress to elucidate many aspects of these unusual hafnia structures.

## REFERENCES

[1] S. Desgreniers and K. Lagarec,"High-density ZrO2 and HfO2: Crystalline structures and equations of state", *Phys. Rev. B*, **59** [13] (1999) 8467.

[2] J.E. Lowther, J.K. Dewhurst, J.M. Leger, J. Haines, "Relative stability of $ZrO_2$ and $HfO_2$ structural phases", *Phys. Rev. B*, **60**[21] (1999) 14485-14487.

[3] R. Ruh and V.A. Patel, "Proposed phase relations in the $HfO_2$-rich portion of the system Hf-HfO_2", *J. Amer. Ceram. Soc.*, **56** [11](1973) 606-607.

[4] A.G. Boganov, V.S. Rudenko, L.P. Makarov, Rentgenograficheskoe issledobanye dviokisei tzirkonya i gafnya pri temperaturah do 2750° (X-ray studies of the dioxides zirconia and hafnia at temperatures up to 2750°), *Dokladi Academyi Nauk*, **160**[5] (1965) 1065-1068. [in Russian].

[5] L. Passerini, "Isomorfismo tra ossidi di metali tetravalenti" (Isomorphism of the oxides of tetravalent metals*)*, *Gaz. Chem. Ital.*, **60** (1930) 762-776. [in Italian].

[6] J.M. Leger, A. Atouf, P.E. Tomaszewki and A.S. Pereira, "Pressure induced phase transitions and volume changes in $HfO_2$ up to 50GPa", *Phys. Rev. B*, **48** (1993) 93-98.

[7] L.Liu, "New high-pressure phases of $ZrO_2$ and $HfO_2$ , *J. Phys. Chem. Solids*, **41** (1980) 331-334.

[8]J. M. Leger, J. Haines and B. Blanzat, "Materials potentially harder than diamond: Quenchable high-pressure phases of transition metal dioxides", *J. Mater. Sci. Let.*, **13** (1994) 1688-1690.

[9]M. Alvisi, S. Scaglione, S. Martelli, A. Rizzo, L. Vasanelli, "Structural and optical modification in hafnium oxide thin films related to the momentum parameter transferred by ion beam assistance", *Thin Solid Films*, **354** (1999) 19-23.

[10]M.Gilo, N. Croitoru, "Study of HfO$_2$ films prepared by ion assisted deposition using a grid-less end-hall ion source", *Thin Solid Films*, **350** (1999) 203-208.

[11]T. Nishide, S. Honda, M. Matsuura, M. Ide, "Surface, structural and optical properties of sol-gel derived HfO$_2$ films", *Thin Solid Films*, **371** (2000) 61-65.

[12] J. Aarik, A. Aidla, A.-A. Kiisler, T. Uustare, V. Sammelselg, "Influence of substrate temperature on atomic layer growth and properties of HfO$_2$ films", *Thin Solid Films*, **340** (1999) 110-116.

[13]B. Andre, L. Poupinet, G.Ravel, "Evaporation and ion assisted deposition of HfO2 coatings: Some key points for high power laser applications", *J.Vac. Sci. Technol.*, *A***18** (2000) 2372-2377.

[14]S. Capone, G. Leo, R. Rella, P. Sicilliano, L. Vasanelli, M. Alvisi, L. Mirenghi, A. Rizzo, "Physical characterization of hafnium oxide thin films and their application as gas sensing devices", *J. Vac. Sci. Technol.*, *A* **16** (1998) 3564-3468.

[15]M. Ritala, M. Leskelä, L. Niinistö, T. Prohaska, G. Friedbacher, M. Grasserbauer, *Thin Solid Films* **250** (1994) 72.

[16] K. Kukli, J. Ihanus, M. Ritala, M. Leskelä, Tailoring the dielectric properties of HfO$_2$-Ta$_2$O$_5$ nanolaminates, *Appl. Phys. Lett.*, **68** (1996) 3737-3739.

[17] J. Aarik, A. Aidla, H. Mändar, T. Uustare, K. Kukli, M. Schuisky, "Phase transformations in hafnium dioxide thin films grown by atomic layer deposition at high temperatures", *Appl. Surf. Sci.* 173 (2001) 15-21

[18] S. Miyake, K. Honda, T. Kohno, Y. Setsuhara, M. Satou and A. Chayahara, "Rutile type TiO$_2$ formation by ion beam dynamic mixing", *J. Vac. Sci. and Technol.*, *A***10** (1992) 3253-3259.

[19] I. Shimizu, M. Kumagai, H. Saito, Y. Setsuhara, Y. Makino, and S. Miyake, Synthesis of CeO$_2$ films by ion beam assisted deposition, in I. Yamada, J. Matsuo and G. Takaoka eds, *Proc. 12$^{th}$ Int. Conf. Ion Implantation Technology, Kyoto, June, 1998*, IEEE, pp943-946

[20] P. Villars and L.D. Calvert, *Pearson's Handbook for Crystallographic Data for Intermetallic Phases*, ASM, Metals Park, Ohio, 1985, Vol. 3, p. 2480.

[21] S. Miyake, I. Shimizu, R. R. Manory, T. Mori, G. Kimmel, Structural Modifications of Hafnium Oxide Films Prepared by Ion Beam Assisted Deposition under High Energy Oxygen Irradiation, *Surf. Coatings Technol.*, in press (2001).

# PLASMA SPRAYED CERAMIC COATINGS ON GLASS AND GLASS CERAMIC SUBSTRATES

R. Gadow,    A. Killinger,    C. Li

Institute for Manufacturing Technologies of Ceramic Components and Composites
University of Stuttgart, Baden-Württemberg,
Allmandring 7b
70569 Stuttgart
Germany

ABSTRACT

Spray coating ceramics on glass substrates is very different from coating on metal substrates as glass material shows exceptional thermophysical properties compared to ordinary metal substrates.

Especially in the case of materials with a low or even negative thermal expansion coefficient like borosilicate glasses or glass ceramics, the thermal spray process requires a very sophisticated temperature guidance and torch kinematic to realise thermally and mechanically stable ceramic glass compounds with a sufficient adhesion of the ceramic coating on the glass substrate. The presentation discusses some aspects of the thermal spray process and the resulting effects on the properties of the acquired glass ceramic compounds.

INTRODUCTION

Plasma sprayed coatings are produced by introduction of powder particles of the material in to a  plasma flame, which melts and propels them towards the substrate. The condition of these particles prior to impact- namely, their temperature, velocity and size – and the condition of the substrate – especially the temperature, chemistry, roughness, etc. – determine the particle spreading and solidification. Depending on these conditions, the resulting splats then have different shape, thickness and contact with the substrate, with important consequences on the deposit microstructure and properties. During deposition the

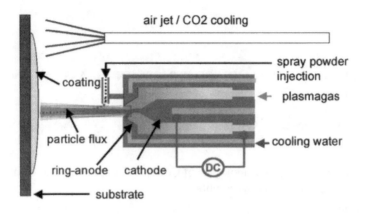

Figure 1. Principle of Atmospheric Plasma Spraying

substrate is heated up by convection (and radiation) of the flame and by heat transfer of the particles that build up the coating (latent heat and lamellae cooling) [1]. In order to reduce the substrate temperature, a simultaneous cooling of the composite is necessary i.e., by $CO_2$ or air gas jet (figure 1).

The application of thermal spray coating on glass was successfully applied on improved ozonizer tubes increasing their efficiency thus cutting down the production costs. The development helps to utilize ozone as an economically competitive alternative to traditional chlorine compounds. In general large quantities of ozone are generated by dielectric barrier discharges (DBD) [2, 3]. Figure 2 shows a borosilicate glass tube with a plasma spray coated metal / ceramic composite coating applied. The thickness of the ceramic coating lies in the region of 600 – 1000 μm.

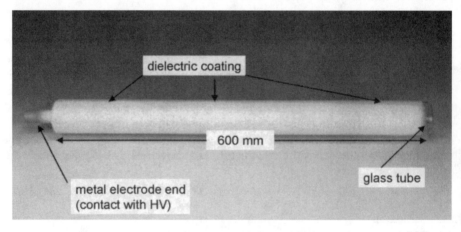

Figure 2. Ozonizer tube with novel coating for the use in commercial ozonizers

A further application of thermal spray coatings on glass is electric isolation on glass ceramic. Because glass ceramic has almost zero or even negative expansion over a wide temperature range and behaves completely insensitive to sudden changes in temperature, it is widely used for heating devices and various domestic glass items, etc. Unfortunately, like other glass materials, glass ceramic also becomes an electrical conductor at high temperatures. This means that glass ceramics cannot serve as an electrical insulator at higher temperatures. However, in newer development heat coils are directly printed on the backside of the glass ceramic. Therefore, it is necessary to apply a dielectric ceramic coating in between to electrically isolate the glass ceramic from the heat coil at elevated temperatures.

GLASS AND GLASS CERAMIC SUBSTRATES

Glass is of considerable importance for the production of scientific, technical, architectural and decorative objects and many articles used in daily life. Glassy materials can be considered as frozen-in liquid, which consist, in the case of oxide materials, of polymer chains with branches and cross linkages. Studies of the melting process and extensive X-ray structural analyses of glass allowed to explain glass as an extended molecular network without symmetry and periodicity [4]. Figure 3 shows the crystalline state of SiO2 (A), the glass network of SiO2 (B) and the glass network of sodium silicate glass (C) represented as two dimensional drawings.

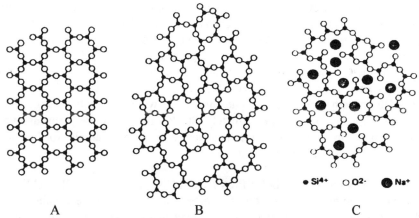

• Si⁴⁺    ○ O²⁻    ⬤ Na⁺

A                    B                    C

Figure 3. Structure of crystallised silica (A), of fused silica (B) and of sodium silicate glass (C)

THERMOPHYSICAL PROPERTIES OF GLASS AND GLASS CERAMIC

At room temperature, glass usually behaves as a solid material. The most important properties of glass as elasticity, rheology and strength depend not only on the molecular mechanism of the deformation process, but on the viscous flow and on the structural peculiarities. To obtain a homogeneous melt of inorganic glass, a temperature is required where the viscosity $\eta$ of the melt is about $10^2$ Poise. The softening point is at $\eta = 10^{7.6}$ Poise. The softening temperature is a function of the rate of heating. For example, for soda lime borosilicate glass softening will occur in the temperature region between 708-815°C if the rate of heating is equal to 0,2 degree/s. Different pieces of glass, each with a different structure, will have different softening temperature when heated at the same rate. For technical applications, it is therefore useful to choose maximum annealing temperature below a temperature (200°C) to prevent unwanted deformations [5].

Inorganic glass is an ionic conductor. The mobility decreases as the ionic radius increases. The glasses are insulators at low temperatures and increase their conductivity as the temperature is raised because of the greater ease with which the ions can move when thermal energy has weakened the binding forces to the silica network. In the temperature range of 25- 1200°C , the resistivity may vary between $10^{19}$ $\Omega$ and 1 $\Omega$.

As already mentioned, all of glass is in the state of undercooled liquid. The main reason for crystallization not taking place for glass then, is mainly due to the fact that the crystal growth, controlled by the diffusion of the components, is far too slow. This is due either to the increasing viscosity of the melt with decreasing temperature, or to the fact that the number of nuclei from which crystallites can form is too low. However, in the case of glass ceramics, the crystallite formation is forced in suitable glass system, in order to obtain materials with special properties. The technical importance of glass ceramics lies in the fact that their properties are determined not solely by the glass percentage, but also decisively by the types of crystals formed. Systems in which crystal phases of very low or even negative thermal expansion are formed have become of great significance. Materials are obtained here with almost zero expansion over a wide temperature range, which are nondeformable up to approximately 800°C and completely insensitive to sudden changes in temperature, and thus can be used, for example, for hot plated and various domestic glass items, etc.

## APPLICATION OF THERMAL SPRAY COATING ON GLASS CERAMIC

Owing to the different properties of glass substrate from metal substrate, the surface of glass substrate will not be grit-blasted as metallic surface before the deposition, because the surface of glass substrate will be damaged by this way and the bonding strength at the interface will be thus decreased. Therefore the bounding mechanism between coating and glass substrate is mainly of chemical nature compared to the ordinary, mainly mechanical bonding mechanisms of

Table 1. Thermophysical properties of investigated glass substrate and coating materials

| material | Young's modulus E [GPa] | Thermal exp. coeff. $\alpha$ [$10^{-6}$ 1/K] | Effective heat capacity Cp [kJ/kg K] | Thermal conductivity $\lambda$ [W/mK] |
|---|---|---|---|---|
| glass-ceramic | 95 | ±0.15 | 0,8 | 1,6 |
| Borosilicate glass | 63 | 3,3 | 0,8 | 1,16 |
| $Al_2O_3$ (APS) | 40-50 | 8,0 | 1,35 (1000°C) 0,78 (30°C) | 5,0 (1000°C) 28,5 (100°C) |
| Mullite (bulk) [6] $3(Al_2O_3)\,2(SiO_2)$ | 145 | 4,3-5,0 | 1,23 (1000°C) 0,77 (30°C) | 3,7 (1000°C) 5,3 (100°C) |
| Cordierite (bulk) [6] $2(MgO)\,2(Al_2O_3)\,5(SiO_2)$ | 40 | 2,2-2,4 | 0,86 (30°C) | 2,1 (1000°C) 2,6 (100°C) |

thermal sprayed coatings. However, in order to achieve a sufficient bonding the glass substrate temperature has to be increased prior to coating.

Recent studies have shown that the deposition temperature has significant influence on the formation of splats after impact. Specifically, at low substrate temperatures the solidification of powder particle can start before the spreading is completed (flower-like splats). Higher temperature usually lead to the formation of thinner, contiguous, disk-like splats. This affects the splat's contact with the substrate through changes in surface tension and wetting, because the proportion of true-contact and no-contact areas has significant influence on the coating properties (e.g. adhesion and quenching stress) as well as the splat formation itself. Better contact leads to faster heat extraction from the splat and thus accelerates the solidification and the diffusion between the substrate and coating material [7,8].

The experimental work of this study is the selection of raw materials and the development of promising coating systems. Four oxide ceramic powders suitable for the APS process are investigated. These are $Al_2O_3$ (type A and B), as well as mullite and cordierite. The latter two materials are of special interest because they have small thermal expansion coefficients getting close to the one of the glass ceramic. For these materials one would expect a good thermophysical compatibility with the glass ceramic substrate.

The quality of the coating is influenced by the combination of spray powder, spray process and spray parameters. Preferentially the used powders should show a narrow grain size distribution and a high flowability. To ensure the quality of the used spray powders and the quality of the coatings, several powder investigations are performed. A SEM picture giving information of the typical shape and porosity of the powders is shown in figure 4.

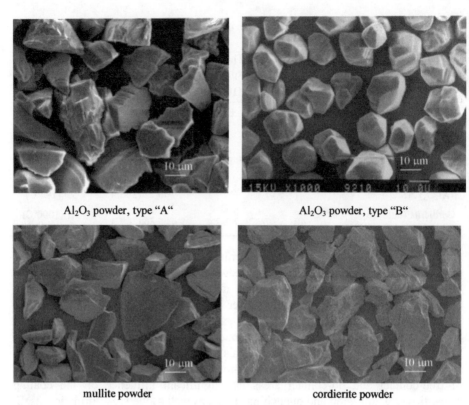

Al$_2$O$_3$ powder, type "A"    Al$_2$O$_3$ powder, type "B"

mullite powder    cordierite powder

Figure 4. The SEM images show typical morphology of the spray powder particles used in this work

All coatings have been deposited onto planar glass samples via atmospheric plasma spraying. Atmospheric Plasma Spraying was performed with a Sulzer-

Metco F4 spray gun with a 6mm Ø nozzle. Powder feeding was performed using a MARK15 screw feeder. The standard spraying parameters are listed below:

| | |
|---|---|
| current (AC): | 600 A |
| gas flow rate: | Ar/H$_2$ 40/14 SLPM (plasma power 41 kW) |
| powder feed rate: | 28 g/min |
| spray distance: | 90 mm |
| preheating temperature: | 320°C |

The movement of spray gun was performed with a meander shape spray pattern. The temperature during the deposition was measured by infrared thermography [9]. In order to achieve high coating temperature (600-700°C) and thus to accelerate the diffusion between the substrate and coating material, air cooling was deactivated during the first deposition step. Because of coating delamination which depends on the coating temperature during the deposition, cooling should be activated since the second deposition step.

## APPLIED THERMAL SPRAY POWDERS

The powders used for this investigation are commercial and custom made powders as well. All powders were extensively analysed prior to spraying, i.e., particle size distribution, particle shape and phase composition were determined by laser optical granulometry, SEM analysis and XRD, respectively.

Several oxide ceramics based on alumina, mullite and cordierite have been investigated so far. Criteria for the selection of thermal spray powders were appropriate thermophysical properties, reduced costs, coating process stability and commercial availability.

$Al_2O_3$ "A", mullite and cordierite powder used for spraying are a fused and crushed quality with a grain size distribution of $-25 + 5$ µm. $Al_2O_3$ "B" is grown from gas phase and is a new kind with spherical shape, excellent flow properties and a very narrow grain size distribution (Fig. 4). Coatings revealed from this powder are usually chemical pure and therefore exhibit higher dielectric breakthrough voltage.

## COATING DEPOSITION AND CHARACTERISATION

The quality of a thermally sprayed coating with regard to microstructure, porosity and other mechanical and electrical properties can be widely varied by tuning the simultaneous spray parameters like supplied energy, substrate preheating and process cooling. Also the physical properties of coating materials (thermal expansion coefficient ($\alpha$), heat capacity ($c_p$) and thermal conductivity ($\lambda$)) are finally influenced by the coating microstructure and porosity.

## OPTICAL MICROGRAPHES

Figure 5 shows the microstructure of the obtained plasma sprayed coatings from the four applied powders introduced in the last section. Same spray conditions have been used for all powders. The coating received from powder $Al_2O_3$ "B" shows significantly lower porosity. As the spray conditions were kept constant for all powders, a reason for that is the extremely narrow grain size distribution of the $Al_2O_3$ "B" powder and hence the more uniform melting of the individual particles. Normally the amount of unmolten particles has the most important impact on the resulting porosity of the coating.

In the microstructure of the mullite and cordierite coatings partially unmolten particles are found. This is because they have very low thermal conductivity and

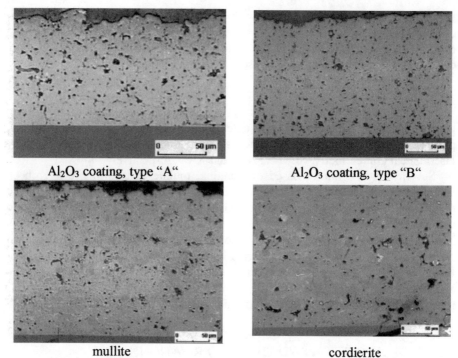

| Al₂O₃ coating, type "A" | Al₂O₃ coating, type "B" |
|---|---|
| mullite | cordierite |

Figure 5. Optical micrographs of the respective microstructures of plasma sprayed coatings using the spray powders from Figure 4

therefore the powder particles were not completely molten in the plasma flame during the flight towards the substrate. A high specific plasma energy or a reduced powder feed rate is necessary to fully melt the particles and achieve a less porous and more homogeneous coating structure. Measured porosities of the plasma sprayed oxide coatings extracted from image analysis on cross section micrographs are summarized in table 2.

Table 2. Measured porosities of plasma sprayed oxide coatings (see figure 5)

| material | porosity p [%] |
|---|---|
| Al₂O₃ "A" | 8,6 |
| Al₂O₃ "B" | 9,6 |
| Mullite | 7,5 |
| Cordierite | 6,6 |

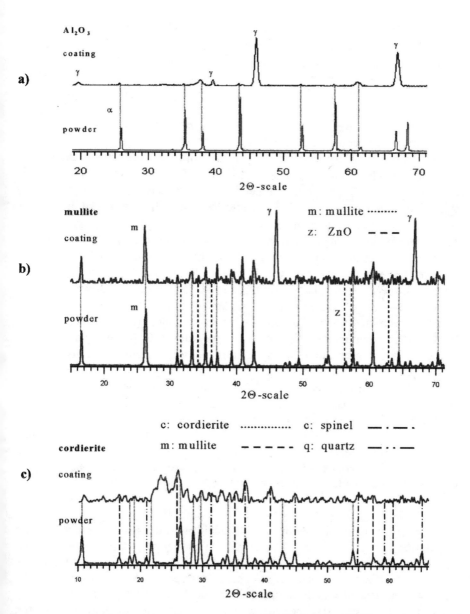

Figure 6. XRD pattern of oxide spray powders and their respective plasma prayed coatings. Strongest peaks are labeled according to JCPDS data base. See text for discussion

## XRD SPECTRA POWDERS / COATING

As already mentioned, plasma sprayed coatings are produced by introduction of powder particles of the material in to a plasma flame, which melts and propels them towards the substrate. Most oxide materials undergo specific phase transformations during the deposition process. Due to the rapid solidification of the particle on the substrate a number of different metastable phases are formed in the coating. These were examined by comparing XRD measurements of the powder and the respective coating. The spectra are shown in Figure 6.

$Al_2O_3$ partially undergoes a phase transition from the hexagonal corundum into the cubic $\gamma$–modification as can be seen in the XRD spectra (Figure 6a). This effect has been extensively investigated in the literature [10, 11]. For many applications this is not appreciated, because the $\gamma$–modification is chemically less stable. Chráska et al. pointed out that the amount of the $\gamma$–phase corresponds to the cooling rate of the molten particles in plasma sprayed coatings [11]. A higher cooling rate results in a higher value for the $\gamma$–phase fraction. Thus, the $\gamma$–phase is a high temperature modification that is formed during the melting process and is frozen in as a metastable phase after rapid solidification on the substrate surface. Strongest peaks are labeled as $\alpha$ and $\gamma$ respectively in Figure 6a.

In the spectrum of the mullite powder the mullite phase (labeled as $m$ in Figure 6b) and some contamination is present, supposedly ZnO (labeled as $z$). In the coating a certain proportion of the mullite phase is decomposed to $\gamma$–$Al_2O_3$ (labeled as $\gamma$).

The XRD spectrum of cordierite shows a rather complex structure (Figure 6c). Besides the major phase ($Mg_2Al_4Si_5O_{18}$, hexagonal, labeled as $c$) also mullite (labeled as $m$), spinel (labeled as $s$) and quartz (labeled as $q$) can be identified.

As cordierite has a pronounced tendency to form amorphous phases when thermally sprayed, clearly identifiable peaks are scarcely detectable in the coating spectrum. Some residuals of cordierite may be visible but the major fraction undergoes transition into an amorphous state.

## SPLAT FORMATION

Since the coating is built-up from individual splats, the splats' morphology has direct influence on the coating's microstructure and properties. As a general trend it could be observed, that with increasing substrate temperature the particles look more smooth, forming round-shaped splats, whereas at lower-temperature the splats become irregular and fragmented. In the latter case the fragments have round edges, suggesting that the separation takes place in a molten or semi-molten state rather than after solidification caused by quenching stress. Figure 7 shows the morphology of splats sprayed on glass ceramic surface at different substrate preheating temperatures. The behaviour can be understood as follows: at no preheating of the glass ceramic substrate (room temperature), the mullite splat has

Innovative Processing and Synthesis of Ceramics

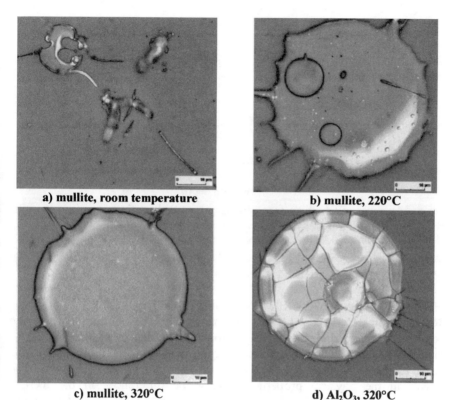

| a) mullite, room temperature | b) mullite, 220°C |
| c) mullite, 320°C | d) Al₂O₃, 320°C |

c) mullite, 320°C  d) Al$_2$O$_3$, 320°C

Figure 7. Typical morphology of mullite and Al$_2$O$_3$ splats on glass ceramic substrate surface at different substrate preheating temperatures

a flattened and splashed shape. Due to the bad wetting ability of the droplet, diffusion between coating and substrate will hardly take place and therefore the coating adheres bad to substrate. For high substrate preheating temperature about 320°C, splats have disk-like shape and the wetting ability of droplet to substrate is good. A good wetting ability will cause a tight contact to the substrate and therefor a higher heat conductivity through the interface. High heat conductivity causes high cooling speed and instant loss of viscosity, therefor the droplet is "frozen" in its initial state without splashing on the surface.

Comparison of mullite and alumina reveals clearly visible structural differences. A distinct micro-crack network can be observed on the Al$_2$O$_3$ droplet (compare Figure 7 c and d). This is because the pronounced difference of the thermal expansion coefficient existing between Al$_2$O$_3$ and the glass ceramic substrate (refer to table 1). As the adhesion to the substrate is strong at high

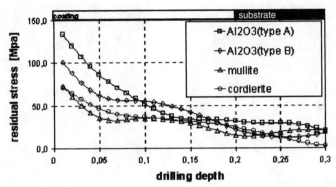

Figure 8. Measured residual stresses of plasma sprayed coatings on glass ceramic, powder sprayed with a lower feed rate (15 g/m)

substrate temperatures, the splat is fixed to the substrate surface and will crack due to the volume shrinkage induced by the cooling.

## RESIDUAL STRESS MEASUREMENTS ON COATED SAMPLES

Residual stresses occur because of macroscopic and microscopic non homogenous elastic and elastic-plastic deformations caused by mechanical and/or thermal load. During the solidification and quenching process a shrinkage of the splats takes place and heat is transferred from the coating to the substrate. The mismatch in the physical properties, mainly the differences in the thermal expansion coefficients (CTE) between coating ($\alpha_c$) and substrate ($\alpha_s$), cause thermal stresses ($\sigma_{th}$) [9].

The residual stresses of coatings were determined by the micro hole drilling method [9]. Figure 8 shows the results of the coatings deposited from the four applied powders with APS. It can be seen that the stresses mainly depend on the CTE of the coating materials. The stress of $Al_2O_3$ "B" coating is smaller than of $Al_2O_3$ "A" coating. This can be explained by its narrow grain size distribution and thus smaller quenching stress occurred during coating deposition.

As already mentioned, phase transformation takes place during the coating deposition of cordierite (predominantly amorphous phase). Amorphous cordierite has a larger CTE (ca. $6 \cdot 10^{-6}$ 1/K ). This explains why the residual stress in the cordierite coating is larger than in the mullite coating.

Glass ceramic has almost zero thermal expansion. The residual stress in the coating after deposition is of tensile nature. Particularly this is not favourable because oxide and glass ceramic have much more compressive than tensile stress. However, it is inevitable during the deposition, because the mismatch in the physical properties between coating and substrate and the quenching stress in the splats are always tensile. Figure 9a shows a macro crack that has formed due to the

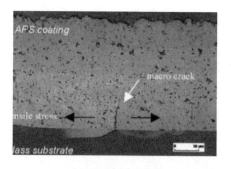

| Al₂O₃ (type A) | cordierite |
| --- | --- |

Al$_2$O$_3$ (type A)

cordierite

Figure 9a. Macro crack caused by tensile stress in the coating near the interface. The crack propagates into the glass substrate.

Figure 9b. Unmolten powder particle in the cordierite coating

tensile stress in the coating near the interface. The crack continues in the glass substrate where it is deflected leading to a spallation within the substrate. In order to prevent those failures, there are several possibilities to suggest:

1)  reduce the preheating temperature of substrate,

2)  reduce the plasma power,

3)  reduce the powder feed rate, and

4)  increase the spray distance.

Because the bonding mechanism between coating and substrate mainly depends on the coating temperature, only the last three ways can be considered. Reducing the

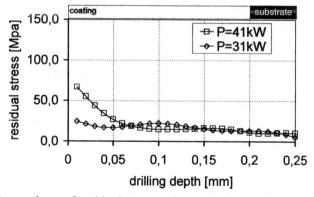

Figure 10. Comparison of residual stresses in cordierite coating on glass ceramic by two different plasma powers (P: plasma power)

Table 3. Measured dielectric breakthrough voltage at 450°C of selected plasma sprayed coatings

| Chemical Composition | Porosity | Dielectric breakthrough voltage at 450°C |
|---|---|---|
| | P [%] | $E_d$ [kV/mm] |
| $Al_2O_3$(type A) | 8,6 | 23,0 |
| $Al_2O_3$(type B) | 9,6 | 25,8 |
| Mullite | 7,5 | 20,9 |
| Cordierite | 6,6 | 9,17 |

plasma power has less influence on residual stress during the coating deposition of $Al_2O_3$, because the CTE of the coating has not significantly changed. The effect is more pronounced for cordierite coating, where unmolten powders can be identified indicating a reduced plasma energy and therefor a reduced thermal mismatch (Figure 9b, 10). There are many other spray parameters that influence the stress situation in the coating. Figure 11 shows the residual stress distribution in the coatings depending on the powder feed rate (Figure 11a) and the spray distance (Figure 11b). An increased spraying distance as well as a lower powder feed rate can reduce the heat transport into the substrate during the deposition process, and therefor lead to a reduction of thermally induced stress in the coating.

ELECTROPHYSICAL PROPERTIES OF DIELECTRIC COATINGS

The dielectric strength has been measured as a function of the coating thickness. Because at room temperature the glass ceramic has also large electrical resistance, specimen (4 mm thickness) must be heated up to 450°C. In table 3 the measured dielectric strengths of the investigated sprayed coatings are listed.

From the investigated materials the powders $Al_2O_3$ (type B) has reached the highest values. Unfortunately, coatings of $Al_2O_3$ possess the largest residual stress. After optimisation of the spray parameters, cordierite features the smallest residual stress values, but the dielectric breakthrough voltage is much less and therefor the coating thickness must be increased. This causes increasing stress again. As a result it can be said that the mullite is the best among the four applied coating material, because it has relatively large dielectric breakthrough voltage and small residual stress, which are required during this coating deposition.

Innovative Processing and Synthesis of Ceramics

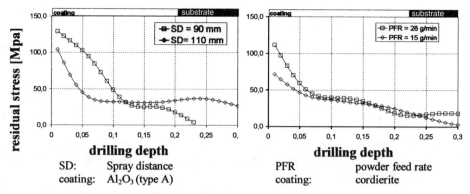

Figure 11. Residual stress distribution in plasma sprayed oxide coatings depending on the powder feed rate and the spray distance

SUMMARY

Phase transformation, morphology of splats, mechanical and electric properties of plasma-sprayed ceramic coatings on glass ceramic were studied. It was found that the preheating of substrate during deposition of glass ceramic is necessary, since bounding mechanism between coating and glass substrate is mainly of chemical nature. However, high preheating temperature causes also large residual stress in the coating. In order to reduce those, substrate cooling must be activated starting the second deposition step. Further spray parameters, which mainly influence the residual stress distribution are plasma power, powder feed rate and spray distance.

REFERENCES

[1]    Pawlowski L; "The Science and Engineering of Thermal spray coatings", John Wiley & Sons, Chichester, New York, Toronto, Singapore, 1995, ISBN 0 471 95253 2

[2]    Gadow R., Riege G., German patent, Nr. 195 11 001.3, 1995

[3]    Friedrich C., Gadow  R. and Killinger, A. "Atmospheric plasma sprayed dielectric coatings for ozonizer tubes", ACerS, 101. Jahrestagung, 25. – 28. April 1999, Indianapolis, USA, Symposium E: Dielectric Materials and Devices

[4]    Zachariasen W.H., J. Am. Chem.Soc., 54 (1932)384 1

[5]    Pulker H.H., "Coatings on Glass", Elsevier,  Amsterdam, Lausanne, New York, Oxford 1999 , ISBN 0 444 50103 7

[6]    Touloukian Y.S., "Thermophysical Properties of Matter" Volume 2, 5,13

[7]     Bianchi L., Blein F., Lucchese P., Grimaud A., Fauchais P., Proc. 7th Nat. Thermal Spray Conf., Boston, USA, 575-579 (1994)

[8]     Kuroda S., Dendo T., Kitahara S., J. Thermal Spray Tech., 4(1), 75-84 (1995)

[9]     Friedrich C., Gadow R., Killinger A., Li C., IR Thermographic Imaging - a Powerful Tool for On-line Process Control of Thermal Spraying, United Thermal Spray Conference and Exposition, Singapore, 2001

[10]    Pawlowski L., „The relationship between structure and dielectric properties in plasma-sprayed alumina coatings", Surface and Coating technology, 35, 285-298, (1988)

[11]    Chráska P., Dubsky J., Neufuss K., Písacka J., „Alumina-Base Plasma-Sprayed Materials Part I: Stability of Alumina and Alumina-Chromia", Journal of Thermal Spray Technologie, 6 (3), 320 - 325, 1997

# Gas Infiltration and Preceramic Polymer Processing

# MANUFACTURING OF BIOMORPHIC SIC-CERAMICS BY GAS PHASE INFILTRATION OF WOOD

Evelina Vogli, Cordt Zollfrank, Heino Sieber and Peter Greil
University of Erlangen – Nuernberg
Department of Materials Science (III)
Glass and Ceramics
Martensstrasse 5 91058
Erlangen/Germany

Joydeb Mukerji
Central Glass and Ceramic Research Institute (CSIR)
65 A N.S.C. Bose ROAD Flat 303 (R) 700 040
Jadavpur
Calcutta/ India

## ABSTRACT

Cellular SiC-ceramic with hierarchically structured micromorphologies are interesting candidates for applications as high-temperature filter, substrate or catalyst support materials. A new approach for processing of microcellular ceramic materials is the reproduction of wood morphologies by biotemplating, where the structural features of the native wood are maintained. Different species of native wood (oak, pine, poplar, bamboo) were converted into carbon preforms by high-temperature pyrolysis at 800°C in argon atmosphere. Subsequent infiltration and reaction with SiO-vapour at 1600°C resulted in a biomorphic β-SiC ceramic. The biodiversity of the natural grown wood offers a large variety of different cellular morphologies and microstructures, which may be useful for creating biomorphic ceramics with improved mechanical and functional properties.

## INTRODUCTION

Porous SiC-ceramics are of increasing interest for applications as catalyst carriers, porous-medium burners, thermally and mechanically loaded light weight structures, as acoustic and heat insulation structures, corrosion resistant materials, biotechnology substrates etc. [1-3]. While previous investigations focused on the

SiC conversion of synthetic carbon materials, recent activities also use biologically derived materials as precursors, e.g. wood or organic fibers. Cellular ceramics with variable pore size and strut thickness can be manufactured from natural materials by biomimetic technologies [4-6].

Wood is a naturally grown composite material with a specific tissue anatomy and exhibits a complex hierarchical cellular structure. The major biopolymeric constituents of wood are cellulose, hemicellulose and lignin, the chemical composition is C (50wt.%), O (44wt.%) and H (6wt%). Wood can be divided into softwoods (coniferous wood or gymnosperms) and hardwoods (deciduous wood or dicotyledonous angiosperms).

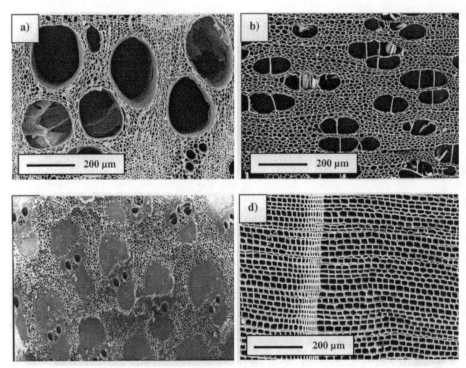

Fig. 1: Cellular morphologies of natural biological structures of a) Oak (quercus robur); b) Poplar (populus nigra); c) Bamboo (cephalsctachyum pergracile); d) Pine (pinus sylvestris).

Coniferous wood is formed predominantly of tracheids with only a small amount of parenchyma, while deciduous wood contains a greater variety of cell types, including thick cell-walled fibres. In contrast to deciduous wood, which exhibits a multimodal pore distribution, coniferous wood has a more uniform morphology and consists of 90-95% tracheids, Fig. 1. The size and wall thickness of the tracheids depends on the growth conditions like water and nutrition supply [7].

Various biotemplating technologies were developed to convert naturally grown structures into ceramic materials. Ota et. al. [8] produced highly porous SiC with wood structure by infiltration of charcoal with TEOS (tetraethyl orthosilicate) and subsequent high-temperature pyrolysis. Greil et. al. [9] converted different kinds of wood by infiltration of pyrolyzed wood preforms with liquid Si into cellular SiSiC ceramic composite with excellent mechanical stability. Biological materials containing silica like rice husks [10] or coconut shells [11] were pyrolysed in $N_2$ atmosphere at 1400°C to obtain SiC. Ohzawa et. al. [12] carbonized cotton-fibre performs, which were infiltrated by a pressure-pulsed chemical vapor technique with $SiCl_4$-$CH_4$-$H_2$ to produce a highly porous SiC ceramic.

In the present work the vapor phase reaction of pyrolysed wood (Oak, Pine, Poplar wood) and Bamboo with SiO has been studied. During the conversion of wood into highly porous biomorphic SiC ceramics, the morphology of the initial wood structure is retained down to the submicrometer length scale. The pyrolysis process, the reaction kinetics during vapour$_{SiO}$ - solid$_{carbon}$ reaction and the final biomorphic cellular SiC-morphology will be discussed.

## EXPERIMENTAL PROCEDURE
Wood derived SiC was produced from different kinds of wood species: decideous wood (Oak-*quercus robur* and Poplar-*populus nigra*), coniferous wood (Pine-*pinus sylvestris*), and Bamboo (*cephalostachyum pergracile*) which differ in densities and porosities (Table 1).

The wood samples were cut perpendicular to the fibre axis. Biocarbon preforms ($C_B$–templates) were prepared by pyrolysing the dried (70°C, 24h) plant specimens in Ar atmosphere. Up to 500°C a slow heating rate of 1K/min was applied, followed by a higher rate of 5K/min up to 800°C. The specimens were hold at the peak temperature for 4h. Details of the experimental set up for the SiO-generation and infiltration of wood derived carbon templates are described in [13].

Table1: *Properties of the processed plant materials.*

| Properties | Oak | Poplar | Bamboo | Pine |
|---|---|---|---|---|
| Porosity [%] | 52 | 73 | 66 | 67 |
| Bulk-Density [g/cm$^3$] | 0,71 | 0,41 | 0,47 | 0,47 |
| **Molecular Composition [wt%]** | | | | |
| Cellulose | 45...49 | 31...60 | 38...43 | 42...52,2 |
| Hemicellulose | 15...21,5 | 15...23 | 19...26 | 8,2...13,4 |
| Lignin | 22...32,2 | 14...24,5 | 25...34 | 26...31,4 |
| Resin/Wax | 1,7...3,2 | 2,3...3,2 | 3...6 | 1...5,9 |
| Fiber Length [mm] | 1...6 | 3...2,1 | 1,4...4,7 | 1,8....4,5 |

The ceramic conversion of the bioorganic specimens was studied by thermal balance analysis (TGA, STA 409, Netzsch Gerätebau, Selb/Germany). The microstructure of the native wood, the pyrolysed char and the SiC processed ceramic was characterised by scanning electron microscopy SEM (XL30, Fa. Philips, NL) and by X ray diffraction, using monochromatic CuK$_\alpha$ radiation (Diffrac 500, Siemens AG, München/Germany). The pore size distribution (down to a pore size of 10 nm) was determined from high pressure Hg porosimetry (Porosimeter 2000, Carlo Erba Instr., Milano/Italy).

## RESULTS AND DISCUSSION

### A.  Pyrolysis of wood

The formation of carbon template structures from wood is a thermally induced decomposition process. Due to the degradation reaction of the biopolymers (cellulose, hemicellulose and lignin), the thermogravimetric curves of wood and Bamboo show a rapid weight loss between 220°C and 320°C in inert atmosphere with a heating rate of 5K/min (Fig. 2). With increasing temperature, dehydration of the biopolymers take place and above 320°C the elimination of volatiles (CO, $CO_2$) occurs. Pyrolytic fragmentation leads to aromatized entities and finally to a highly crosslinked carbon with graphite like crystalline structure at temperature above 1200°C [14]. The C$_B$ - template shows a final weight loss ranging from 70wt.% (Oak) to about 80wt.% (Bamboo) [15].

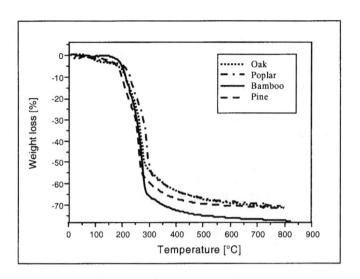

*Fig. 2:* Weight loss of different kinds of wood and bamboo during pyrolysis in inert atmosphere.

Table 2: *Properties of the carbon templates.*

| Properties | | Oak | Poplar | Bamboo | Pine |
|---|---|---|---|---|---|
| **Porosity [%]** | | 69 | 75 | 73 | 76 |
| **Density [g/cm³]** | | | | | |
| | Bulk | 0,41 | 0,23 | 0,31 | 0,34 |
| | Skeleton | 1,31 | 0,9 | 1,16 | 1,41 |
| **Shrinkage [%]** | | | | | |
| | Axial | 18,7 | 21,5 | 19,9 | 22 |
| | Radial | 28 | 29,6 | 27,8 | 28 |
| | Tangential | 34,8 | 36 | 28,3 | 30 |
| **Specific Surface Area [m²/g]** | | 89 | - | 38 | 46 |

Despite the weight loss during the pyrolytic conversion and a pronounced linear shrinkage in axial, radial and tangential direction (Table 2), the microstructural features of the cellular wood and bamboo are reproduced with high precision in the carbon template. The large specific surface area is due to a porous strut morphology of the biocarbon template in the submicrometer range.

## B.    SiO gas – Carbon template reaction

The overall reactions of SiO-gas phase with the wood derived carbon template may be explained by:

$$SiO_g + 2C_s \rightarrow SiC_s + CO_g \qquad (1)$$
$$2SiO_g + 3C_s \rightarrow 2SiC_s + CO_{2g} \qquad (2)$$

FTIR analysis during the reaction at temperatures between 1200-1600°C shows, that with increasing reaction temperature the $CO/CO_2$ ratio in the exhaust gas increased. Depending on the processing temperature SiC-ceramic with different porosities can be produced [16, 17]. According to reaction (1) and (2) only half or two third of the initial carbon volume is converted to SiC resulting in the formation of a porous SiC reaction product with a high surface area.

Table 3: *Porosity and specific surface area of wood derived SiC.*

| Properties | Oak | Poplar | Bamboo | Pine |
|---|---|---|---|---|
| **Porosity [%]** | 73 | 91 | 68 | 86 |
| **Specific Surface Area [m²/g]** | 9,8 | 14,4 | 8,9 | 16,1 |

The specific volume change associated with the reaction of the wood derived carbon and the SiO-gas to SiC depends on the initial char density $\rho_C$:

$$\Delta V/V_C = (V_{SiC} - V_C)/V_C = \frac{n_{SiC} M_{SiC}\, \rho_C}{n_C\, M_C\, \rho_{SiC}} - 1 \qquad (3)$$

where n, M and $\rho$ denote the number of moles, the molecular weight and the densities of the phases indicated. The density of SiC ($\rho_{SiC}$) is $3{,}21 g/cm^3$. The density of the wood char after pyrolysis at 800°C ($\rho_C$) is in the range of $0{,}9...1{,}41 g/cm^3$ (Table 2). Thus, the formation of SiC by SiO-gas infiltration results in a decrease of the overall volume in the range 25-40% (Eq. 3), (Table 3).

The cellular anatomy of the biological specimen was maintained in the char as well as in the biomorphic SiC-ceramic (Figs. 1, 3 and 4).

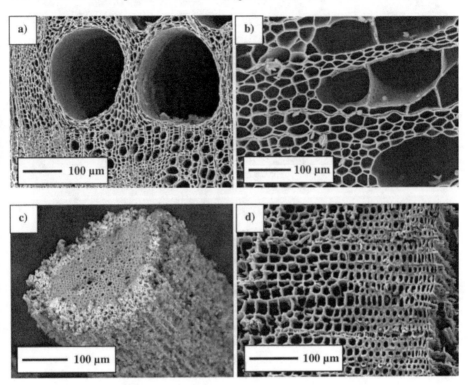

*Fig. 3: SiC derived from a) Oak; b) Poplar; c) Bamboo; d) Pine, after SiO infiltration at 1600°C.*

During the SiO-gas infiltration into the carbon template ß-SiC was formed as the major phase with minor amounts of α-SiC. With increasing reaction temperature α-SiC transformed into ß-SiC and samples infiltrated at 1600°C for 16h revealed ß-SiC only [17]. The grain size of ß-SiC ranges from 200 to 400 nm.

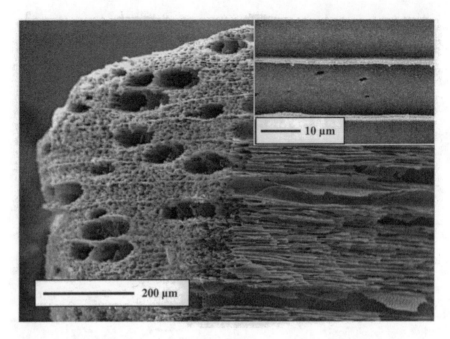

*Fig. 4: Poplar derived SiC-ceramic (infiltrated at 1600°C for 8 hours).*

## CONCLUSIONS

Highly porous SiC-ceramics with a large specific surface area can be prepared by SiO–gas infiltration of carbonised wood. After the solid – gas reaction the microstructure of the native wood is retained in the SiC-ceramics down to the micrometer length scale. While the conversion of deciduous wood yields microcellular SiC-ceramics with a multimodal cell diameter distribution from a few micrometer to 100µm, infiltration of coniferous wood species results in a nearly monomodal cell distribution with a mean cell diameter between 10 and 30µm.

## ACKNOWLEDGEMENT

Financial support from the Volkswagen Foundation under contract # I/73 043 is gratefully acknowledged.

# REFERENCES

[1]O. Pickenäcker, K. Pickenäcker, K. Wawrzinek, D. Trimis, W. E. C. Pritzkow, C. Müller, P. Goedtke, U. Papenburg, J. Adler, G. Standke, H. Heymer, W. Tauscher, F. Jansen, "Innovative ceramic materials for porous-medium burners", *Interceram, * **48** [5&6] 1-12 (1999).

[2]M. J. Ledeoux, S. Hantzer, C. P. Huu, J. Guille, M. P. Desaneaux, "New synthesis and uses of high specific surface SiC as a catalytic support that is chemically inert and has high thermal resistance", *Journal of Catalysis,* **114** 176-185 (1988).

[3]K. Okada, H. Kato, K. Nakajima, "Preparation of silicon carbide fiber from activated carbon fiber and gaseous silicon monoxide", *Journal of American Ceramic Society,* **77** 1691 (1994).

[4]H. Sieber, C. Hoffmann, A. Kaindl, P. Greil, "Biomorphic cellular ceramics", *Advanced Engineering Material,* **2** [3] 105-109 (2000).

[5]Y. Ohzawa, M. Yoshimura, K. Nakane, K. Sugiyama, "Preparation of high-temperature filter by partial densification carbonised cotton wool with SiC", *Materials Science and Engineering,* **A242** 26-31 (1998).

[6]D. W. Shin, S. S. Park, "Silicon/Silicon carbide composites fabricated by infiltration of a silicon melt into charcoal", *Journal of American Ceramic Society,* **82** 3251-3253 (1999).

[7]R. Wagenführ, *Holzatlas;* Carl Hanser Verlag, München Wien (2000).

[8]T. Ota, M. Takahashi, T. Hibi, M. Ozawa, S. Suzuki, Y. Hikichi, "Biomimetic process for producing SiC "Wood"", *Journal of American Ceramic Society,* **78** [12] 3409-3411 (1995).

[9]P. Greil, T. Lifka, A. Kaindl, "Biomorphic silicon carbide ceramics from wood I and II", *Journal of European Ceramic Society,* **18** 1961-1983 (1998).

[10]G. Padmaja, G. P. Mukunda, "Production of SiC from rice husks", *Journal of American Ceramic Society,* **82** [69] 1393-1400 (1999).

[11]A. Selvan, N. G. Nair, P. Singh, "Synthesis and characterisation of SiC whisker from coconut shells", *Journal of Material Science Letters,* **17** 57 (1998).

[12]Y. Ohzawa, A. Sadanaka, K. Sugijama, "Preparation of gas-permeable SiC shape by pressure-pulsed chemical vapour infiltration into carbonised cotton-cloth preforms", *Journal of Materials Science,* **33** 1211-1216 (1998).

[13]E. Vogli, J. Mukerji, C. Hoffmann, R. Kladny, H. Sieber, P. Greil, "Conversion of Oak to cellular silicon carbide ceramic by vapour phase reaction with SiO", *Journal of American Ceramic Society,* (2001) in print.

[14]A. Oberlin, "Carbonization and graphitization", *Carbon,* **22** [6] 521-541 (1984).

[15]A. Kaindl, "Cellular SiC ceramics from wood", Ph.D. thesis, University of Erlangen-Nuremberg (2000) Germany.

[16]J. C. Lee, M. J. Park, "Effect of hold time on reaction of silicon monoxide with activated carbon fiber composites", *Carbon,* **37** 1075-1080 (1999).

[17]E. Vogli, J. Mukerji, H. Sieber, P. Greil, "Biomorphic SiC ceramic produced by gas infiltration of wood", Proccedings of Materials Week 2000, Symposium: Biomimetic Processing of Structural Material, Munich/Germany, September 2000, in press.

# MANUFACTURING OF ANISOTROPIC CERAMICS FROM PRECERAMIC POLYMER INFILTRATED WOOD

Cordt Zollfrank, Ralf Kladny, Heino Sieber and Peter Greil
Department of Materials Science (III) Glass and Ceramics, University of Erlangen-Nuernberg, D-91058 Erlangen, Germany

Günter Motz
Institute for Materials Research, University of Bayreuth, D-95440 Bayreuth, Germany

**ABSTRACT**

Light-weight cellular SiOC-ceramics were manufactured by vacuum infiltration of hardwood and softwood with two different organosilicon polymers, polymethylphenylvinylsilsesquioxane (PMPVS) and polymethylhydrosiloxane (PMHS). The infiltrated polymers were cured by heat-treatment at 220°C and pyrolysed at 800°C in inert atmosphere. Infiltration, reaction, curing and pyrolysis of the organosilicon infiltrated wood was monitored by Fourier Transform Infrared spectroscopy (FTIR), transmission electron microscopy (TEM), thermogravimetrical analysis (TGA), and scanning electron microscopy (SEM). After pyrolysis of the infiltrated wood samples at 800°C an amorphous SiOC-glass/carbon composite was formed.

**INTRODUCTION**

Biotemplating, a novel technology of biomimetic processing, involves material synthesis from biologically grown materials into ceramic composites by fast high-temperature processing. In the last years, various biotemplating high-temperature techniques were developed to convert biological materials into porous ceramics [1]. Considerable efforts have been devoted to the production of biomorphic SiC ceramics from wood and organic fibres. Si coated cellulose fibres [2] and $Si_3N_4$ coated cotton fibers [3] were converted into SiC fibers by annealing in Ar-atmosphere at 1200-1600°C. The infiltration of different kinds of pyrolysed wood preforms with liquid Si yields the formation of SiSiC composites [4,5]. Shin and Park [6] used charcoal for fabrication of biomorphic SiSiC-composites in a similar process. Vogli et al. demonstrated that pyrolysed wood can be converted to a highly porous cellular SiC ceramic by vapor phase reaction with gaseous SiO

[7]. Ota et al. produced biomorphic ceramics by infiltration of native wood and charcoal with metal-organic solutions, e.g. tetraethyl orthosilicate (TEOS) or titanium tetra-isopropoxide (TTiP) [8,9]. After high-temperature processing the wood structures were converted into porous SiC or $TiO_2$ ceramics.

The liquid infiltration and chemical modification of the wood constituents (lignin, cellulose, polyoses) with organosilicon polymers and subsequent conversion into SiOC/C-ceramic composites is a novel approach presented in this paper. The preceramic polymers contain reactive groups like Si-H-bonds which can be linked to the OH-functional groups on the surface of the cell wall of the biopolymers. The preceramic polymers bonded to the cell wall result in a significant reduction of shrinkage and weight loss beneficial for the wood-to-ceramic conversion process.

**EXPERIMENTAL PROCEDURES**

Hardwood (beech, poplar) and softwood (white fir, pine) were infiltrated under a vacuum with liquid organosilicon polymers PMPVS (Wacker, Germany) and PMHS (Fluka, Germany). Infiltration was repeated 3 times to obtain a maximum in weight percentage gain. Infiltration experiments were performed without the use of solvents. The PMPVS already contained a curing catalyst (Pt-catalyst). Temperature induced condensation reaction (165-220°C) of PMHS to the hydroxyl function of wood polymers was confirmed by FTIR measurements.

Table 1: *Charateristic of the infiltrated polymers.*

| Polymer | Formula | Mol. Weight [g/mol] | Molecular structure | Viscosity (20°C) [mPa·s] |
|---------|---------|---------------------|---------------------|--------------------------|
| PMPVS | $RSiO_{1.5}$; <br><br> $R = (C_6H_5)_{2.8}(CH_3)_{1.5}(CH=CH_2)H$ | 1000 - 1300 | branched | 1400 |
| PMHS | $(CH_3)_3SiO[CH_3HSiO]_{25-35}Si(CH_3)_3$ | 1500-1900 | linear | 15-40 |

The wood samples (2 g) were dried at 120°C for 15 h. After infiltration the organosilicon polymers were cured at 220°C for 4 h. Weight gain (WG) upon infiltration was calculated according to equation 1. The pyrolysis was performed in a quartz tube furnace which was continuously flushed with nitrogen. The furnace was heated at a rate of 1K/min from room temperature to 400°C and from 400 to 800°C at rate of 5K/min. The peak temperature was held for 4 h. After cooling down to room temperature, the weight loss (WL) was calculated according to equation 2. Volume changes (VC [%]) was determined according to equation 3.

$$WG = (m_{inf} - m_o) / m_o \cdot x \, 100 \, [\%] \qquad (1)$$

$$WL = (m_{pyr} - m_{inf}) / m_{inf} \, x \, 100 \, [\%] \qquad (2)$$

$$VC_{ax} = (V_{pyr} - V_o) / V_o \, x \, 100 \, [\%] \qquad (3)$$

$m_{inf}$: mass of MHS-infiltrated sample, $m_o$: mass of native sample. $m_{pyr}$: mass of pyrolysed sample (Eq. 2). $V_{pyr}$: Volume of the pyrolysed samples, $V_o$: Volume of the infiltrated wood samples (Eq. 3).

FTIR spectra were recorded on a Nicolet Impact 420-T instrument operated in the directed reflection mode. 1.024 scans at a resolution of 4 cm$^{-1}$ were taken to obtain one spectrum with an acceptable signal to noise ratio. Spectra were obtained from the outer surface of the infiltrated wood samples. Infiltration was proved by axial cleaving the samples and recording a spectrum from the inner surface. TGA measurements were performed with a Netzsch STA 409 instrument in the temperature range from 20°C to 1000°C at a rate of 5 K/min (N$_2$-atmosphere). X-ray diffraction diagrams were obtained from the pyrolysed and reacted samples using a Siemens Diffrac 500 diffractometer using monochromated (CuK$_\alpha$) radiation. SEM micrographs were obtained from a scanning electron microscope (XL30, Philips) operated at 25 kV. TEM studies were performed on ultra-thin (Reichert ultra-microtome) sections of embedded (Spurr medium) specimens in a Zeiss EM 10C microscope operated at 60 kV.

## RESULTS AND DISCUSSION
*Infiltration of solid wood with organosilicon polymers*
The weight gain depends on the initial density of the infiltrated wood, Fig. 1. Softwoods generally exhibit a smaller weight gain after infiltration compared to hardwoods. Hardwoods show a larger variety of the geometrical density (polplar: 0.3 g/cm$^3$; beech: 0.6 g/cm$^3$), and the size of the vessels (20-70 μm) are larger compared to the lumen of the tracheids of softwoods (10-20 μm).

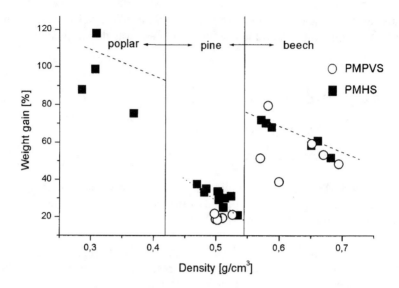

<u>Figure 1:</u>     *Weight gain after infiltration of native wood with organosilicon polymers as a function of wood density.*

PMHS and PMPVS entirely infiltrated the solid wood sample as proved by FTIR spectroscopy. After infiltration and curing of the infiltrated polymers no volume change of the wood samples could be measured. The linear PMHS yields an average weight gain of 78% for the infiltration of hardwood and 30% for softwood. The branched PMPVS shows about 5% less weight gain after infiltration.

TEM studies in Fig. 2 of the infiltrated wood samples show that the cell walls were coated with a thin layer of the organosilicon polymer. At the cell corners polymer deposits are observed. Small cavities like the functional elements of the solid wood sample (e.g. pit holes) and some of the cell lumen are completely filled with polymer. However, the organosilicon polymers do not cause a swelling of the wood cell wall. Thus, internal cell walls like the secondary wall 2 (S2, cellulose-rich) are not infiltrated and chemically modified by the organosilicon polymers [10]. The silane groups present in PMPVS and PMHS react with the hydroxyl function of the wood polymers only at the cell wall surface. Condensation reaction depicted in Fig. 3 of PMHS and PMPVS was proved by FTIR spectroscopy.

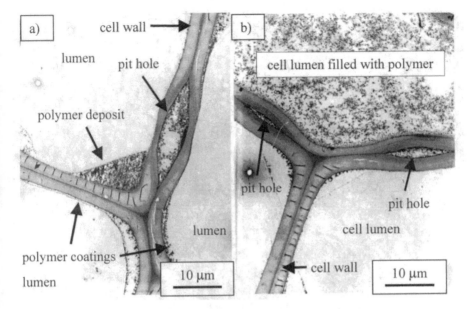

Figure 2: *TEM images of white fir early wood infiltrated with PMHS: a) coated cell walls; b) completely infiltrated cell lumen.*

$$R\text{-OH} + \underset{*}{\overset{*}{H-\underset{\underset{*}{\overset{|}{O}}}{\overset{|}{Si}}-CH_3}} \longrightarrow R\text{-O}-\underset{\underset{*}{\overset{|}{O}}}{\overset{|}{Si}}-CH_3 + H_2$$

Figure 3: *Condensation reaction of silane functionalised polymers with the hydroxyl function of wood polymers (R-OH, e.g. R = lignin).*

*Pyrolysis*

TGA measurements show that the weight loss for native and infiltrated wood started at 220°C as depicted in Fig. 4. A maximum weight loss is determined between 260 and 360°C. Char yield for native wood is 78 %. The weight loss decreased by 28% for the PMPVS and PMHS infiltrated wood.

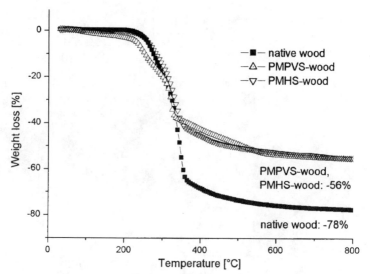

<u>Figure 4:</u>    *Weight loss of organosilicon polymer infiltrated poplar wood during heating in $N_2$-atmosphere.*

After pyrolysis of the infiltrated wood samples the total shrinkage is reduced, Fig. 5.

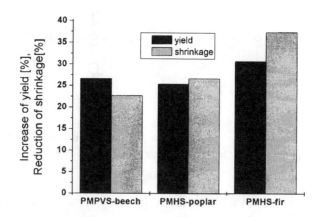

<u>Figure 5:</u>    *Increase of char yield and reduction of shrinkage of the infiltrated samples after pyrolysis at 800°C in inert atmosphere compared to native wood.*

Innovative Processing and Synthesis of Ceramics

SEM micrographs reveal, that the cellular structure of wood is retained after pyrolysis at 800°C. The dark grey phase represents the former cell walls (carbon-phase), whereas the light grey phase is the pyrolysed polymer, Fig. 6. The polymer infiltrated wood samples are converted into biomorphic SiOC-glass/carbon composites.

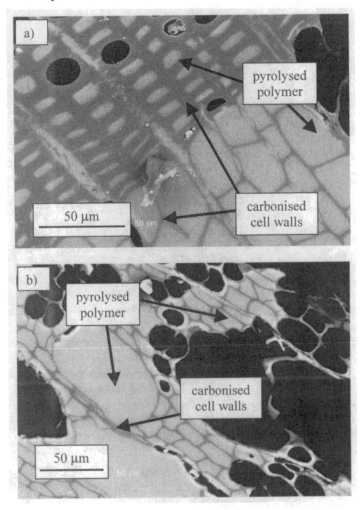

Figure 6: *SEM-micograph (backscattered electrons mode) of infiltrated and pyrolysed samples at 800°C: a) white fir; b) poplar.*

## CONCLUSIONS

Cellular SiOC/C-ceramics can be manufactured by infiltration of native wood with organosilicon polymers. Linear organopolysiloxanes (PMHS) exhibit a higher weight gain after infiltration compared to branched organopoly-silsesquioxanes (PMPVS). The covalent bonding between the organosilicon polymers and the biological matrix improves the ceramic yield during pyrolysis and reduces the shrinkage compared to native wood samples.

## ACKNOWLEDGEMENT

The authors are grateful to VW-Foundation (grant I/73 043). The authors appreciate the technical assistance of Mrs. G. Friedrichs and Dr. K.-P. Gaffal (Institute of Pharmaceutical Biology, FAU-Erlangen) for preparing the TEM-samples and the reproduction of the micrographs.

## REFERENCES

[1] H. Sieber, C. Hoffmann, A. Kaindl, and P. Greil, "Biomorphic Cellular Ceramics," *Adv. Eng. Mat.*, **2** 105 (2000)

[2] R.V. Krishnaro, and Y.R. Mahajan, "Preparation of Silicon Carbide Fibres from Cotton Fibre and Silicon Nitride," *J. Mat. Sci. Lett.*, **15** 232 (1996)

[3] H. Sieber, H. Friedrich, A. Kaindl, and P. Greil, "Crystallization of SiC on biological precursors," *Proc. of the 101st Annual Meeting of the ACerS*, Bioceramics: Materials and Application III, Ceramic Transactions **110**, 81 (2000)

[4] P. Greil, T. Lifka, and A. Kaindl, "Biomorphic Silicon Carbide Ceramics from Wood: I. and II.," *J. Europ. Ceram. Soc.*, **18** 1961 (1998)

[5] C.E. Byrne, and D.E. Nagle,"Cellulose derived composites – A new method for material processing," *Mat. Res. Innovat.*, **1** 137 (1997)

[6] D.W. Shin, and S.S. Park, "Silicon/Silicon Carbide Composites Fabricated by Infiltration of a Silicon Melt into Charcoal," *J. Am. Ceram. Soc.*, **82** 3251 (1999)

[7] E. Vogli, J. Mukerij, C. Hoffmann. R. Kladny, H. Sieber, and P. Greil, "Conversion of Oak to cellular silicon carbide by vapor phase reaction with SiO," *J. Am. Ceram. Soc.*, in print (2001)

[8] T. Ota, M. Takahashi, T. Hibi, M. Ozawa, S. Suzuki, Y. Hikichi, and H. Suzuki, "Biomimetic Process for Producing SiC-Wood," *J. Am. Ceram. Soc.*, **78** 3409 (1995)

[9] T. Ota, M. Imaeda, H. Takase, M. Kobayashi, N. Kinoshita, T. Hirashita, H. Miyazaki, and Y. Hikichi, "Porous Titania Ceramics prepared by mimicking silicified wood," *J. Am. Ceram. Soc.*, **83** 1521 (2000)

[10] R.M. Rowell, "Chemical Modification of Wood," *Forest Products Abstracts*, **6**, 363 (1983)

# Rheological Behavior

# MODELING THE DYNAMIC RHEOLOGICAL BEHAVIOR OF AGAR BASED AQUEOUS BINDERS

K. C. Labropoulos*, D. E. Niesz* and S. C. Danforth*

Rutgers University
Malcolm G. McLaren Center for Ceramic Research
Department of Ceramic and Materials Engineering
607 Taylor Road
Piscataway, NJ 08854

## Abstract

This study proposes a fundamental microstructural/rheological model for agar gels that describes their dynamic rheological behavior during gelation. The concept of a sigmoidal type temperature dependence of the monomeric friction coefficient, $\zeta_o$ , of the agar molecule is introduced to describe the increase of relaxation times of the agar molecules with decreasing temperature. At high temperatures (close to 373 K), the model behaves like a dilute solution of flexible polymers. At low temperatures (close to 273 K) contributions from an equilibrium modulus term are predominant. Dynamic rheology was employed to study the applicability of this model for agar gels. Gelation curves were obtained experimentally for various agar concentrations and cooling rates were fitted successfully to the theoretical model predictions.

## I. Introduction

Powder injection molding (PIM) is a low cost and high quality forming method for the fabrication of complex shaped parts. Typically, the binder removal step limits the shape complexity and size of the injection molded parts that can be made without failure.[1] However, high solids loading (>60 vol%) mixtures with agar gel based aqueous binders have been successfully used for injection molding.[2-4] The main advantage of these binders is that they are environmentally benign and exhibit thermoplastic rheological behavior. Also, since the binder phase consists mainly of water (97-99 vol%), binder removal is greatly simplified.

---

* Member, American Ceramic Society
Sponsored by the Malcolm G. McLaren Center for Ceramic Research, Rutgers University and Honeywell, PowderFlo™ Technologies, NJ

Parts made employing these binders are typically dried overnight under ambient conditions. The use of these binders may allow the fabrication of ceramic or metal parts with large cross sections, as well as extensive variations in thickness, within one part.

Agar is a gel-forming polysaccharide, that is typically extracted from red seaweeds (Gracilaria and Gelidium).[5-7] Agar gels are prepared by dissolution of agar powder in boiling water and then cooling to a temperature below the gelation point (typically < 40°C). During cooling, the homogeneous sol changes gradually into an elastic and transparent gel network. This sol-gel transition is not only reversible, but also path dependent. The formation of the three-dimensional gel network has been extensively studied, and the double-helix model is widely accepted.[5-8] The basic structural unit of the gel network is the "double helix" in which each agar molecular chain (with an average molecular weight estimated to be 120,000) forms a left-handed, 3-fold helix of 1.90 nm pitch and is translated axially, relative to its partner, by 0.95 nm.[5-8] The association of chains is temperature dependent. At temperatures significantly above the gelation point (about 40°C), agar molecules adopt a random coil conformation. Below the gelation temperature, double helices are formed that are interconnected so as to build the three-dimensional network. The double helices are normally aggregated to higher order assemblies, termed suprafibers, which contain 10 to $10^4$ double helices and are responsible for most of the strength of the gel. Upon reheating, the three-dimensional gel network is destroyed. However, agar gels exhibit a significant hysteresis between the gelation and liquefaction temperatures.

In the past, the gelation behavior of these binders has been studied using dynamic rheology.[9-13] Nevertheless, to the best of our knowledge, no fundamental rheological model exists that describes the rheological behavior of agar gels throughout the temperature range they encounter during injection molding. This study proposes a model, based on the theories for dilute and crosslinked polymers.

## II. Development of a Rheological Model for Agar Gels

### Rheological Behavior of Agar Sols in the High Temperature Range

At high temperatures, the molecules should exist in a random coil conformation, and thus the proposed model should reduce to the form predicted by the Rouse theory (equations 1-3).[14] In this theory, the flexible polymer molecule is divided into N equal submolecules.[14] The chain molecule is represented by $N$ segments (each containing q monomers) joining $N+1$ identical beads with complete flexibility at each bead. The Rouse theory predicts:[14]

$$[G']_R = \sum_{p=1}^{N} \omega^2 \tau_p^2 / (1 + \omega^2 \tau_p^2) \qquad (1)$$

$$[G'']_R = \sum_{p=1}^{N} \omega\tau_p \Big/ (1+\omega^2\tau_p^2) \qquad (2)$$

$$\tau_p = \frac{\sigma^2 N^2 f_o}{6\pi^2 p^2 kT} = \frac{\alpha^2 P^2 \zeta_o}{6\pi^2 p^2 kT} = \frac{S^2 P \zeta_o}{6\pi^2 p^2 kT} \qquad (3)$$

where $[G']_R$ and $[G'']_R$ are the reduced intrinsic storage and reduced intrinsic loss modulus respectively, $\tau_p$ is a relaxation time, $\zeta_o$ is the monomeric friction coefficient, $S$ is the mean square separation of the ends of a submolecule, $T$ is the temperature, $\omega$ is the frequency of the applied oscillatory deformation and $p$ is an index number. More details regarding derivation of these equations can be found in the original literature.[14]

### Temperature Dependence of the of the Relaxation Times

As the temperature is lowered from the *'dissolution temperature'* to a value below $T_{gel}$, the relaxation times of the contributing viscoelastic entities should increase. The temperature dependence of the relaxation times is expected to be complex. However, based on the shape of the curve of the temperature dependence of $G'$ and $G''$ of agar gels[9-11] we propose that the monomeric friction coefficient for agar changes with temperature in the following manner:

$$\zeta_o(T) = \zeta_{HT} + \frac{\zeta_{LT} - \zeta_{HT}}{1 + \exp\left(\dfrac{T-T_o}{b_o}\right)} \qquad (4)$$

At high temperatures, $\zeta_o(T)$ takes the value $\zeta_{HT}$, whereas at low temperatures the limit is $\zeta_{LT}$. The transition temperature range is characterized by the temperature $T_o$, and the slope $b_o$, which is related to the temperature width of this transition. For simplicity, any contributions from the temperature dependence of $S$ can be lumped into the $\zeta_o(T)$ term. With these assumptions, the relaxation times $\tau_p$ can then be calculated using the Rouse theory as follows:[14]

$$\tau_p = \frac{\tau_1^{ref} T_{ref}}{p^2 T} \left[ 1 + \frac{r_\zeta - 1}{1 + \exp\left(\dfrac{T-T_o}{b_o}\right)} \right] \qquad (5)$$

where $\tau_1^{ref} = \dfrac{S_{ref}^2 P \zeta_{HT}}{6\pi^2 kT_{ref}}$ , is the terminal relaxation time of the random coils at the arbitrary reference temperature $T_{ref}$ in the high temperature range (example, $T_{ref}$ = 370 K) and $r_\zeta = \zeta_{LT}/\zeta_{HT}$ .

### Gelation Equation

In this study, $T_{gel}$ is defined as the temperature during cooling of an agar sol, above which any associated agar molecules cannot form a continuous three-dimensional network. When the temperature drops below $T_{gel}$, there exists a driving force for gelation proportional to the degree of undercooling, ($\Delta T = T_{gel} - T$); the higher the value of $\Delta T$ the more enhanced the kinetics at the initial stages of gelation.[11] At any given time until the end of the associating processes, the system consists of the solvent and two 'fractions' of agar molecules. The first fraction contains those agar molecules that are not associated with any others. The rheological behavior of these 'free' agar molecules is theorized to be described by the adapted Rouse model (equations 1-5). The second fraction contains the remaining agar molecules that are associated with at least another partner, forming helices and suprafibers and/or physical entanglements. The rheological behavior of this fraction will be described by an adaptation of the model for crosslinked polymers, as developed by Mooney.[15]

We can now define, $n_{free}$ as the number of non-associated agar molecules per cm³, $n_{assoc}$ as the number of associated molecules per cm³, $n_{total}$ as the total number of agar molecules per cm³ and $t = 0$ as the moment during cooling when $T = T_{gel}$. At $t > 0$, the rate of association, $dn_{assoc}/dt$, should depend on the available concentration of non-associated molecules, $n_{free}$. The rate should also be affected by the presence of the agar network that has already been developed. To a first approximation, the rate of association should be proportional to the number of associated molecules, $n_{assoc}$ per unit volume. We can then approximate the gelation equation as follows:

$$\frac{dn_{assoc}}{dt} = \xi(T) \cdot n_{assoc} \cdot n_{free} = \xi(T) \cdot n_{assoc} \cdot (n_{total} - n_{assoc}) = r \cdot n_{assoc} \cdot \left(1 - \frac{n_{assoc}}{n_{total}}\right)$$

(6)

where $r = \xi(T) \cdot n_{total}$ is the net intrinsic association rate and is dependent on temperature. The above equation has the form of the Verhulst or logistics equation and can be solved subject to the initial condition $0 < n_{assoc}^o < n_{total}$ :

$$n_{assoc} = \frac{n_{assoc}^{o} \cdot n_{total}}{n_{assoc}^{o} + (n_{total} - n_{assoc}^{o}) \cdot e^{-rt}} \qquad (7)$$

The temperature dependence of the net intrinsic association rate $r(T)$ is proposed to have the following empirical form:

$$r(T) = A_1 \cdot \left( \frac{T_{gel} - T}{T_{gel} - T_{low}} \right)^{A_2} \qquad T_{low} < T < T_{ref} \qquad (8)$$

where $A_1$, $A_2$ are constants to be calculated. The significance of $T_{low}$ is that it defines the lower limit of applicability of the above equation. The value of $T_{low}$ is typically the mold temperature applied in PIM utilizing these binders.

### Network Contributions

When the temperature is lowered well below $T_{gel}$ and after infinite time, the dynamic moduli are essentially independent of frequency.[9,11,16-18] As indicated in the literature[11,17,19] there exists a $c^x$ dependence of the asymptotic equilibrium modulus, $G_e$, where $x$ is a power typically higher than 2 and $c$ is the agar concentration in $g/cm^3$. $G_e$ is obtained at low temperatures and after very long time. However, at the early stages of gelation, the network contributions are dependent on frequency.

In fact, at the early stages of gelation, contributions to the dynamic moduli should be expected from both fractions of agar molecules: the associated agar molecules that form the network and the non-associated agar molecules. Regarding the network contributions, it should be recognized that during the initial stages of gelation, the network is rather flexible. The flexibility of the network influences the frequency dependence of the network contributions to the dynamic moduli of the system. With increasing rigidity of the network, its relaxation times are shifted progressively to higher values and thus the network becomes less frequency dependent.

To account for the unique properties of agar gels, Mooney's theory for cross-linked polymers is adapted to include an equilibrium dynamic loss modulus term, $G''_e$. In particular, this addresses the different mechanisms of energy storage and dissipation in agar gels at low temperatures, which are probably associated with a combination of bending and stretching of the semi-rigid structural units (suprafibers).[20] It is proposed that, to a first approximation, the frequency dependence is proportional to $\omega^u$, where u is a fitting parameter:

$$u = \frac{1}{2}\left(1 - \frac{n_{assoc}}{n_{total}}\right) \tag{9}$$

The network contributions to the dynamic moduli can thus be approximated:

for $n^o{}_{assoc} < n_{assoc} < \dfrac{c_{AB} \cdot 10^4 N_A}{M}$

$$G'\big|^t_{network} = \delta_{SA} \cdot \left(\frac{n_{assoc} \cdot c}{n_{total}}\right)^{\kappa} + n_{assoc}kT \cdot \delta_3 \omega^u \tag{10}$$

$$G''\big|^t_{network} = \delta_{VA} \cdot \left(\frac{n_{assoc} \cdot c}{n_{total}}\right)^{\lambda} + n_{assoc}kT \cdot \delta_4 \omega^u \tag{11}$$

whereas for $\dfrac{c_{AB} \cdot 10^4 N_A}{M} < n_{assoc} < n_{total}$

$$G'\big|^t_{network} = \delta_{SB} \cdot \left(\frac{n_{assoc} \cdot c}{n_{total}}\right)^{2} + n_{assoc}kT \cdot \delta_3 \omega^u \tag{12}$$

$$G''\big|^t_{network} = \delta_{VB} \cdot \left(\frac{n_{assoc} \cdot c}{n_{total}}\right)^{2} + n_{assoc}kT \cdot \delta_4 \omega^u \tag{13}$$

$$\delta_{SA} = \delta_{SB} \cdot c_{AB}{}^{2-\kappa} \tag{14}$$

$$\delta_{VA} = \delta_{VB} \cdot c_{AB}{}^{2-\lambda} \tag{15}$$

where $c$ is the agar concentration in wt%, $\kappa$ is the exponent of the first term of $G'\big|^t_{network}$, $\lambda$ is the exponent of the first term of $G''\big|^t_{network}$, $c_{AB}$ is the concentration in wt% above which the exponents $\kappa$ and $\lambda$ of the first terms in equations 10 and 11 respectively change to a value of 2, and $\delta_{SA}$, $\delta_{SB}$, $\delta_{VA}$, $\delta_{VB}$ are microstructure related constants. For continuity of equations 10-13 the parameters $\delta_{SA}$, $\delta_{SB}$ and $\delta_{VA}$, $\delta_{VB}$ are related by equations 14 and 15 respectively.

## IV. Experimental Procedure

The agar powder (NF S-100) used to prepare agar gels was provided by Frutarom Meer Corp. North Bergen, NJ. The powder was dissolved in boiling deionised and distilled water (pH 6.5). The heated sol was then held at 100°C for 3 hours and was continually stirred. No water losses occurred during the preparation of the gels. The sol was then allowed to cool to room temperature. Table I lists the compositions prepared in this study.

**Table I.** Composition of Agar Gels

| Material | Designation | | |
|----------|------|------|------|
|          | AG1  | AG2  | AG3  |
| Deionized water (wt%) | 99.0 | 98.0 | 97.0 |
| Agar (wt%) | 1.0 | 2.0 | 3.0 |

Dynamic rheological measurements were performed on an Advanced Rheometric Expansion System (ARES) controlled strain dynamic rheometer (Rheometric Scientific Inc. Piscataway, NJ). The geometry used was parallel plate (25 mm diameter). The samples were cut into a disk (height $\approx$ 2.0 mm, diameter 25 mm) from bulk pieces of gels. The samples were then allowed to relax for at least two hours in sealed water-saturated containers before testing. The linear viscoelastic region (LVER) of the gels was determined by strain sweep experiments at a frequency $\omega$=100 rad/s, at various temperatures. Generally, strains below 5% were within the LVER for most of the temperatures. A strain ($\gamma$) of 1%, was used for all the dynamic measurements for all compositions and temperatures studied. Time sweep experiments were used to study the gelation kinetics of the agar gels. The gel was initially melted at 95°C for 1 min before subsequent cooling. During the experiments, the temperature varied at an instrument-preset rate.

**V. Results and Discussion**

The typical gelation behavior of 1 wt% agar gels is shown in Fig. 1, where the storage modulus, $G'$, is plotted vs. temperature during cooling from 90°C, at two cooling rates (12°C/min and 36°C/min). Moduli values that correspond to torque values below the rheometer resolution (approximately 10 Pa) are not shown. Gel formation is signified by the steep rise in $G'$, which reaches a maximum value of approximately $10^4$ Pa. The corresponding values of the maximum storage moduli of 2 and 3 wt% agar gels are approximately $4.5 \cdot 10^4$ Pa and $10^5$ Pa respectively. These values are in good agreement with a $c^2$ dependence for agar concentrations at and above 1 wt%.[1,17,19]

A decrease in the slope of the $G'$ vs. $T$ curves is evident with increasing cooling rate. The relationship of the slope with the association rate will be discussed later. The maximum values for $G''$ were approximately an order of magnitude lower than $G'$. The temperature dependence of the storage or loss moduli of higher agar concentrations (2 wt% and 3 wt%) was similar to Fig. 1, however, the curves were shifted to higher temperatures.

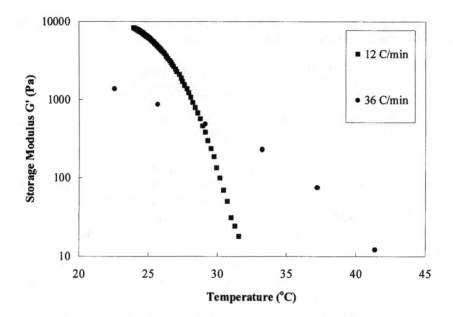

**Figure 1.** Storage modulus, G', vs. temperature showing gelation behavior of 1 wt% agar gels on cooling from 90°C for two cooling rates as indicated. $\gamma = 1\%$, $\omega = 100$ rad/s, 25mm parallel disk geometry.

In order to simplify the fitting of the model, certain parameters were estimated based on available information in the literature. Specifically, the terminal relaxation time, $\tau_l^{ref}$, at the selected reference temperature $T_{ref} = 373.15$ K, was estimated using the following equation, which was assumed to be applicable even at such high temperatures:[14,20]

$$\tau_1^{ref} = \frac{6[\eta]\eta_s M}{\pi^2 R T_{ref}} \tag{16}$$

where $[\eta]$ is the intrinsic viscosity of the sol, $\eta_s$ is the viscosity of the solvent and M is the average molecular weight of agar (approximately 120,000). Estimation of the $\tau_l^{ref}$ using the above equation yields a value in the order of $10^{-4}$ to $10^{-5}$ s. The values $T_o = 303.15$ K, $b_o = 10$ K, $r_\zeta = \zeta_{LT}/\zeta_{HT} = 100$ were chosen based on initial fitting attempts. The value of $r_\zeta$ was estimated using data at the initial onset of gelation (slightly above $T_{100}$). A value of $r_\zeta = 100$ was selected as representative for all samples.

The critical gelling concentration, $c_o$, (typically 0.05 wt%) [7,18] below which no gelling can occur, was used to estimate the value of $n^o_{assoc}$. The low temperature limit of applicability for the proposed model, $T_{low}$, was taken as 283.15 K which is the value of typical mold temperature for PIM utilizing these binders.[2] The values of $A_1$ and $T_{gel}$ were studied as a function of the average cooling rate between 90°C and 50°C, as will be discussed later. However, the exponent $A_2$ was given an empirical value of 0.2 for all gel compositions. Initial fitting tests showed that this value was a good compromise between providing a temperature dependence to the net intrinsic association rate and allowing the same value to be used for a wide range of cooling rates and agar concentrations.

Regarding the network contributions to the dynamic moduli, the exponents $\kappa$ and $\lambda$ were taken as 5 and 2 respectively. Again these values were chosen largely empirically so that the proposed model fits to a wide range of cooling rates and agar concentrations. However, the agar concentration parameter $c_{AB}$ was allowed to vary with cooling rate. A value of 0.1 was selected for both $\delta_3$ and $\delta_4$ using data at the onset of the gelation, in the vicinity of the $G'$-$G''$ crossover. The parameters $\delta_{SB}$ and $\delta_{VB}$ were assumed to have the values of 11104 Pa and 361 Pa respectively, for all gel compositions and cooling rates. These values were calculated so that the theoretical maximum storage and viscous moduli for the 3 wt% agar gels were $10^5$ Pa and $3.25 \cdot 10^3$ Pa respectively. These were the typical values for the dynamic moduli of agar gels for this agar concentration. The values of $\delta_{SA}$ and $\delta_{VA}$ were calculated using equations 14,15.

The proposed model was then fitted to a series of experimental gelation curves over a large range of cooling rates and agar concentrations. The fitting was limited to data above the instrument resolution and up to the maximum values of the storage modulus. The average relative error between the theoretical and experimental values of $G'$ or $G''$ over the fitted range was typically below 15%. Figure 2 illustrates the typical accuracy of fitting. The results for the cooling rate dependence of the net intrinsic association rate, $A_1$ and the agar concentration parameter $c_{AB}$ are presented in Table II. The cooling rate dependence of $T_{gel}$ is shown in Figure 3.

Table II indicates that the net intrinsic association rate, $A_1$ increases with increasing cooling rates. Also, to a first approximation, $A_1$ is proportional to agar concentration, $c$. It is theorized that as the temperature drops faster, the agar molecules have a higher probability of association, although probably they associate in a more random manner. The agar concentration parameter $c_{AB}$ decreases with increasing cooling rate. This behavior probably indicates an increased degree of randomness (and thus the dynamic moduli will tend towards a $c^2$ type concentration dependence) as the cooling rate is increased.

**Table II.** Net Intrinsic Association rate, $A_1$ , and Agar Concentration Parameter $c_{AB}$ as a Function of Cooling Rate

| | | Cooling Rate (°C/min) | | | |
|---|---|---|---|---|---|
| | | 1 | 12 | 24 | 36 |
| $A_1$ (x10,000) | AG1 | 48 | 66 | 350 | 378 |
| | AG2 | | 156 | 579 | |
| | AG3 | 110 | 221 | | 649 |
| $c_{AB}$ (wt%) | AG1 | 0.57 | 0.44 | 0.28 | 0.25 |
| | AG2 | | 0.45 | 0.28 | |
| | AG3 | 0.69 | 0.27 | | 0.28 |

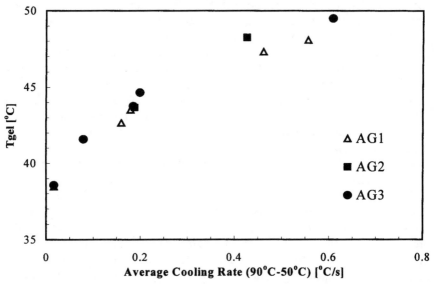

**Figure 2.** Plot of storage modulus (G' : closed circles) and viscous modulus (G" : open squares) as a function of temperature for a 3 wt% agar gel cooled from 90°C at 12°C/min. Solid and dashed lines indicate the fitting of the proposed model to G' and G" respectively, with parameters as described in the text. $\gamma = 1\%$, $\omega = 100$ rad/s, 25mm parallel disk geometry.

Figure 3 is a plot of the theoretical gelation temperature (as defined for use with this model) vs. the average cooling rate from 90°C to 50°C. As can be seen, $T_{gel}$ is approximately independent of agar concentration, at least in the concentration range examined in this study. This indicates that the driving force for network formation, $\Delta T = T_{gel} - T$ is only dependent on the temperature and the thermal history of the sol and is independent of the number of agar molecules available, at least in the agar concentration range studied here. It should be noticed that this behavior may seem to contradict the concentration dependence of the apparent gelation temperature as measured by other techniques. In particular, the apparent $T_{gel}$ is often obtained by dynamic rheological measurements as the temperature at which an arbitrary value of storage modulus is reached during cooling.[9,11,13,17,19] Alternatively, spectroscopic methods are employed, in which case the apparent $T_{gel}$ is defined as the temperature during cooling of a gel where the optical rotation changes rapidly.[5,11,17] Both approaches indicate a concentration dependence of the apparent $T_{gel}$. However, the definition of the theoretical $T_{gel}$ as given in this study, allows for association of agar molecules to take place before any macroscopic technique (rheology or spectroscopy) can detect it. Therefore, it is probably closer to the actual concentration dependence of the true gelation temperature.

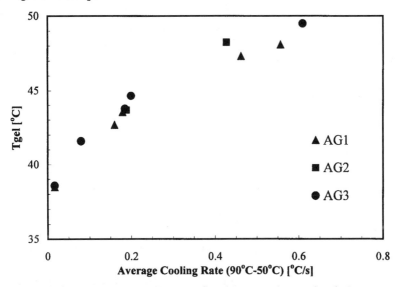

**Figure 3.** Gelation temperature, $T_{gel}$ , as fitted to experimental gelation curves vs. average cooling rate (from 90°C to 50°C).

The concentration dependence of the net association rate, $r(T)$, is governed by the concentration dependence of the net intrinsic association rate, $A_1$, where it was found to be approximately proportional to agar concentration (Table II).

$T_{gel}$ increases with increasing cooling rate. However, outside the range of cooling rates achieved in this study, it is not apparent whether an upper limit of $T_{gel}$ exists. For very low cooling rate, $T_{gel}$ seems to have a value of 38°C. In general, the average cooling rate between 90°C and 50°C provides a robust measure of the effect of the thermal history on the gelation behavior of agar gels. For any given cooling rate in the range studied here and by using Table II and Fig. 3 one can estimate the values of $A_1$, $c_{AB}$ and $T_{gel}$. Then, using the proposed model with the assumptions discussed earlier, a very good prediction of the evolution of the dynamic moduli vs. time and temperature can be obtained.

The time-temperature range over which gelation curves were studied, covers to a large degree the practical time-temperature range for forming methods such as powder injection molding and gelcasting. In such applications, knowledge of the rheological behavior of the binder phase is very important, since it is the binder phase that imparts the fluidity to the particle/binder mixture. Although this study focused on developing a fundamental dynamic rheological model for agar gels containing no ceramic or metal powders, the proposed model will form the basis for the development of a model describing the rheological behavior of gel/particles mixtures. Such a model will be an invaluable tool in optimizing the forming parameters as a function of the size and shape of a part. Therefore, the full potential of these novel binders will be reached and the applicability of powder injection molding and gelcasting will be significantly enhanced.

**Acknowledgments**    We thank the Malcolm G. McLaren, Center for Ceramic Research and Honeywell, PowderFlo™ Technologies, for financial support.

**References**
[1]    F. F. Lange *J. Am. Ceram. Soc.,* **72** [1] 3-15 (1989)
[2]    R. P. Rohrbach, J. R. Schollmeyer and A. J. Fanelli U.S. Patent No 5,397,520
[3]    A. J. Fanelli, R. D. Silvers, W. S. Frei, J. V. Burlew and G. B. Marsh *J. Am. Ceram. Soc.* 72 [10] 1833-1836 (1989)
[4]    T. Zhang, S. Blackburn, J. Bridgwater *British Cer. Trans.* **93** [6] 229-233 (1994)
[5]    I. C. M. Dea, A. A. McKinnon and D. A. Rees, *J. Mol. Biol.,* **68** 153-172 (1972)
[6]    S. Arnott, A. Fulmer, W. E. Scott, I. C. M. Dea, R. Moorhouse and D. A. Rees *J. Mol. Biol.,* **90**, 269-284 (1974)
[7]    P. Harris *"Food Gels"* Elsevier, New York (1990)
[8]    D. A. Rees *Chemistry and Industry,* 630-636 (1972)
[9]    K. C. Labropoulos, S. Rangarajan, D. E. Niesz and S. C. Danforth *J. Am. Ceram. Soc.,* (in print)
[10]    K. C. Labropoulos, S. Rangarajan, D. E. Niesz and S. C. Danforth *Proceedings 102nd Annual Meeting of the Am. Ceram. Soc.,* St. Louis, MO, April 30, 2000 (Innovative Processing and Synthesis of Ceramics, Glasses and Composites, Paper No. C4-066-00)

[11] Z. H. Mohammed, M. W. N. Hember, R. K. Richardson and E. R. Morris *Carbohydr. Polym.* **36** 15-26 (1998)

[12] I. T. Norton, D. A. Jarvis and T. J. Foster *Int. J. Biol. Macromolecules*, **26** 255-261 (1999)

[13] M.-F. Lai and C.-Y Lii *Int. J. Biol. Macromolecules*, **21**, 123-130 (1997)

[14] P. E. Rouse, Jr. *J. Chem. Phys.*, **21** 1272-1280 (1953)

[15] M. Mooney *J. Polym. Sci.*, **34** 599-626 (1956)

[16] M. Manno, A. Emanuele, V. Mantorana, D. Bulone, P. L. San Biagio, M. B. Palma-Vittoreli and M. U. Plama *Phys. Rev. E*, **59** 2222-2230 (1999)

[17] R. Lapasin and S. Pricl *"Rheology of Industrial Polysaccharides: Theory and Applications"* Blackie A & P, London (1995)

[18] Z. H. Mohammed, M. W. N. Hember, R. K. Richardson and E. R. Morris *Carbohyd. Polym.*, **36** 27-36 (1998)

[19] V. Normand, D. L. Lootens, E. Amici, K. P. Plucknett and P. Aymard *Biomacromolecules*, **1** 730-738 (2000)

[20] J. D. Ferry, *"Viscoelastic Properties of Polymers"* $3^{rd}$ Ed. Wiley, NY, (1980)

# Mechanical Alloying

# EFFECTS OF ALLOYING WITH Nb, Re, AND Al ON YIELD STRENGTH OF MoSi$_2$

Adel A. Sharif*
University of Michigan, Flint, Engineering Science
303 East Kearsley Street, 213 MSB, Flint, Michigan 48502

Amit Misra, John J. Petrovic**, and Terence E. Mitchell**
Materials Science and Technology Division, Los Alamos National Laboratory
MS K765, Los Alamos, NM 87545.

ABSTRACT

The effects of alloying MoSi$_2$ with Re, Al, Nb, and their combinations within the solubility limit in C11$_b$ structure on the yield strength were investigated from room temperature to 1600°C. The yield stress behavior shows three distinct regions, solid solution softening, constant yield stress, and solid solution hardening with increasing temperature Solid solution softening was observed below ≈600°C due to alloying with Al and Nb, constant yield stress with varying temperature and alloying elements occurred at ≈600-800°C, and anomalously rapid solid solution hardening at ≳800°C was observed by alloying with Nb and Re. The results are explained by observations on the operative slip systems and their effects on the yield stress.

INTRODUCTION

Molybdenum disilicide (MoSi$_2$) is a potential candidate to replace superalloys as a structural material for high-temperature applications. MoSi$_2$ has superior oxidation resistance, higher melting point, and lower density in comparison with superalloys but inferior low temperature (<900°C) ductility. Furthermore, MoSi$_2$ exhibits a low strength at high temperatures (>1200 °C). For example, below ≈900 °C, the fracture toughness of MoSi$_2$ is in the range of 2-4 MPa$\sqrt{m}$ [1], and

Research was sponsored by the US Department of Energy, Office of Basic Science, Division of Materials Science.
* Member, American Ceramic Society
** Fellow, American Ceramic Society

the 0.2% offset yield strength of MoSi$_2$ at 1600°C is about 20 MPa [2]. The problem with pest oxidation of MoSi$_2$ in the temperature range of 500 to 800°C may be alleviated by alloying [3,4].

A unit cell of the body centered tetragonal C11$_b$ structure of MoSi$_2$ is shown in Fig. 1. Single crystals of MoSi$_2$ can be deformed plastically in compression down to ambient temperatures in all major crystallographic directions except the [001] orientation [5]. However, polycrystalline MoSi$_2$ is brittle at temperatures below ≈900°C due to lack of sufficient number of slip systems. Alloying or reinforcing with a second phase may lower the brittle-to-ductile transition temperature (BDTT) of MoSi$_2$. However, ductile-phase toughening with metallic phases has limited applicability in MoSi$_2$ due to the chemical reaction with silicon to form silicides, and reinforcing with ceramic second phases such as SiC and ZrO$_2$ has only a modest effect on enhancing plastic flow and increasing toughness [6].

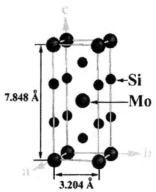

Fig. 1. Unit cell of the body-centered tetragonal C11$_b$ structure of MoSi$_2$.

First principles calculations indicate that alloying of MoSi$_2$ while maintaining its body-centered tetragonal (C11$_b$) structure may result in improved mechanical properties [7]. For example, Al and Nb may enhance ductility and Re may increase strength [7]. Improvements in both low and high temperature mechanical properties of MoSi$_2$ have been reported by alloying MoSi$_2$ with small amounts of Al, Nb, and Re (<2 at%) [2,8-11]. During alloying, below the solubility limits of alloying elements in the C11$_b$ structure of MoSi$_2$, Al substitutes for Si, whereas Re and Nb substitute for Mo. The solubility limits of Re, Nb and Al in MoSi$_2$ have been reported as ≈2.5 at%, ≈1.3 at% and ≈2.7 at%, respectively [12]. Although improvements in the ambient temperature toughness have been reported by alloying of MoSi$_2$ beyond the solubility limits with Nb and Al [13,14], the rates of improvement per fraction of solute are not as considerable as those observed in single phase alloys. Furthermore, the presence of secondary phases with a lower high-temperature strength than the matrix alloy would degrade the mechanical properties at high temperature (>1500°C) for which MoSi$_2$ is the best candidate [15,16].

The present investigation explores the possibility of improving mechanical properties of MoSi$_2$ both at ambient temperatures and high temperatures

Innovative Processing and Synthesis of Ceramics

concurrently. The high temperature improvements in the mechanical properties are focused on improving the yield strength and the low temperature investigations are concerned with increasing ductility. Since ductility and yield strength do not go hand in hand and an increase in one in general results in a decrease in the other, microalloying techniques are employed to take advantage of anomalous solid solution hardening and softening which has been observed in the aforementioned results to obtain concurrently enhanced room temperature (RT) ductility and high temperature strength in single-phase polycrystalline $MoSi_2$.

## EXPERIMENTAL PROCEDURE

Unalloyed $MoSi_2$, ternary $(Mo,Re)Si_2$, $Mo(Al,Si)_2$, $(Mo,Nb)Si_2$, quaternary $(Mo, Re)(Al,Si)_2$ and $(Mo,Nb)(Al,Si)_2$, alloys were prepared by arc-melting elemental Mo, Re, Si, Nb, and Al in an argon atmosphere. The amounts of the added solutes were 2.5, 1, and 2 at% for Re, Nb, and Al, respectively, for all alloying compositions except for the quaternary $(Mo, Re)(Al,Si)_2$ alloy which contained 1 at% Re. Each sample was turned over and remelted 4 times to ensure homogeneity. The samples were polished with SiC paper up to 4000 grit and finished with 0.05 $\mu$m colloidal silica. A Micromet 4 (Buehler, IL) was used for RT microhardness testing with a 1000 g load. Compression testing was performed on $2 \times 2 \times 4$ mm$^3$ samples in air using an Instron 1125 machine at an initial strain rate of $\approx 1 \times 10^{-4}$ s$^{-1}$. Samples were characterized by optical microscopy, transmission electron microscopy (TEM) and scanning electron microscopy (SEM). Phase identification was done by energy dispersive spectroscopy (EDS) in SEM.

## RESULTS

The effects on yield strength of $MoSi_2$ due to alloying with Re, Al, and Nb, within the solubility limit in C11$_b$ structure, may be summarized as:

(a) There was no significant solution softening or solution hardening in $MoSi_2$ at the temperature range of about 600-800°C due to any aforementioned alloying element or their combinations. This temperature range will be called constant yield stress temperature range (CYSTR) henceforth.

(b) Alloying $MoSi_2$ with Re resulted in solid solution hardening above CYSTR, hence, an increase in the yield stress and raising of the brittle-to-ductile transition temperature (BDTT).

(c) Alloying $MoSi_2$ with Al and Nb resulted in concurrent solid solution softening below CYSTR, i.e. a decrease in the yield stress, and solid solution strengthening above CYSTR, i.e. an increase in yield stress.

The compilation of the results of compression testing for samples of unalloyed MoSi$_2$ and those containing 2.5 at% Re, 2 at% Al, and 1 at% Re + 2 at% Al are shown in Fig. 2. Since unalloyed MoSi$_2$ would not yield prior to brittle fracture below 900°C, low temperature data for yield strength of unalloyed MoSi$_2$ are estimated from the results of single crystal compression testes [5,17] and are shown as a dashed line in the plot.

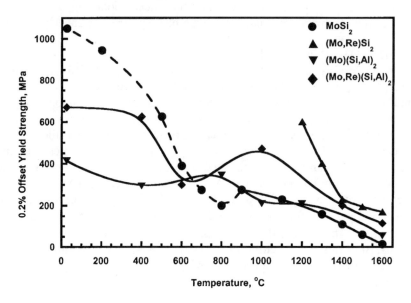

Fig. 2. Effects of alloying with 2.5 at% Re, 2 at% Al, and 1 at% Re + 2 at% Al on yield strength of MoSi$_2$. Low temperature data for yield strength of unalloyed MoSi$_2$ are estimated from the results of single crystal compression tests [5,17].

Yield strength of unalloyed polycrystalline MoSi$_2$ was very low (14 MPa) at 1600°C and increased to 276 MPa by lowering the compression testing temperature to 900°C. The RT yield strength of MoSi$_2$, estimated from the single crystal data was about 1050 MPa. The addition of 2 at% Re increased the yield strength of MoSi$_2$ by a factor of ≈12 to 170 MPa at 1600°C and raised the BDTT to 1200°C. The addition of 2 at% Al to MoSi$_2$ had a significant effect on lowering the BDTT to ≤25°C while increasing the yield strength at 1600°C by a factor of ≈4 to 55 MPa. Among alloying elements investigated here, 2.5 at% Re and 2 at% Al were most effective in increasing strength at 1600°C and improving RT ductility, respectively.

Addition of 1 at% Re+2 at% Al to MoSi$_2$ combined the beneficial effects of both alloying elements and resulted in an enhanced ambient temperature compressive plasticity and high temperature strength compared to the unalloyed samples. However, the improvement in the room temperature plasticity was less than that of samples alloyed with 2 at% Al alone as evident from the value of the room temperature yield strength of (Mo, 1 at% Re)(Si, 2 at% Al)$_2$ alloy, 670 MPa. Similarly, the enhancement in the high temperature strength was less than that of only Re containing samples but greater than that of only Al containing samples.

The effects of 1 at% Nb as an alloying element by itself in lowering BDTT and enhancing the high temperature strength of MoSi$_2$ was comparable to the combined effects of 1 at% Re and 2 at% Al (Fig. 3). The room temperature yield strength of (Mo, 1 at% Nb)Si$_2$, 500 MPa, was higher than that of 2 at% Al containing samples and lower than that of (Mo, 1 at% Re)(Si, 2 at% Al)$_2$ samples. The 0.2% offset yield strength of (Mo, 1 at% Nb)Si$_2$ samples at 1600°C, 143 MPa, was an order of magnitude greater than that of unalloyed MoSi$_2$. Samples containing 1 at% Nb+2 at% Al did not exhibit any improvements in the mechanical properties and were excluded from further considerations.

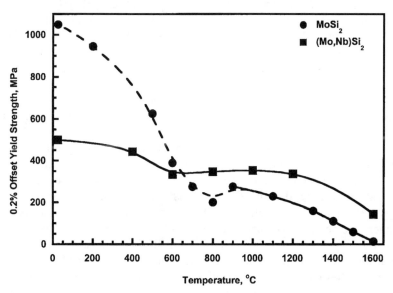

Fig. 3. Effects of alloying with 1 at% Nb on the yield strength of MoSi$_2$.

## DISCUSSIONS

### Solid Solution Hardening and Softening

The mechanisms of solid solution hardening and softening in MoSi$_2$ in the presence of Re, Al, and Nb are discussed elsewhere [2,11,18,19]. The anomalously high solid solution hardening rates observed in the presence of Re may be attributed to the interactions between the dislocations and the elliptical strain fields around these point defect complexes generated due to the creation of a Si vacancy by addition of every four Re atoms to MoSi$_2$ [20]. Alloying with Re was found to change the deformation mechanism from a climb-controlled dislocation mechanism with primarily ⟨100] slip in unalloyed MoSi$_2$ to a viscous glide-controlled dislocation motion with primarily ⟨100] and some 1/2⟨111] slip above 1200°C.

The hardening effects observed in samples containing 2 at% Al and 1 at% Nb are greater than those which could be attributed to solid solution hardening in metals due to size misfit and elastic moduli mismatch alone as discussed previously for the effects of Re in MoSi$_2$ [2]. Deformation of MoSi$_2$+1 at% Nb at 1200°C resulted in dislocations with predominantly ⟨100] Burgers vectors [18]. ⟨110] type sessile dislocations were formed from the [100]+[010]=[110] reaction. The dislocation substructures were consistent with the increased glide resistance due to Nb alloying. The point defect structures in (Mo,Nb)Si$_2$ and Mo(Si,Al)$_2$ alloys need to be studied in more detail to interpret the high hardening rates at elevated temperatures.

Both 2 at% Al and 1 at% Nb individually resulted in low-temperature (<600°C) solid solution softening and high-temperature (>1300°C) solid solution hardening of MoSi$_2$. However, when MoSi$_2$ was alloyed with both elements, their beneficial effects of each at low and high temperatures were nullified by the presence of the other. Two phenomena may explain these observations. First, Al and Nb may react during arc melting resulting in an Al-Nb intermetallic phase, thus, eliminating the beneficial effects of either element alone. No evidence of an Al-Nb phase was found during EDS investigations in SEM. Second, the effects of alloying elements may be nullified if the size misfit effects on the matrix due to one element are opposite to those of the other element. Similar effects have been observed by alloying Al with solutes of different atomic radii [21].

Regarding the solution softening, the tetragonal structure of MoSi$_2$ is closely related to the hexagonal structure and easy transformation from tetragonal to hexagonal is possible. It is known that the 1/2⟨111] dislocations dissociate into two co-linear 1/4⟨111] partials separated by a stacking fault on {110) [22]. The stacking sequence in the fault is ABCABC, which is also the stacking sequence in the C40 structure. Thus, the alloying elements that stabilize the C40 structure with respect to $C11_b$ are likely to lower the stacking fault energy and thereby increase the width of the faulted region. This change in the stacking fault energy surface ($\gamma$-surface) by alloying is likely to affect the Peierls stress and dislocation nucleation and mobility [10]. At low temperatures ($\leq 600$ °C) where softening is observed, dislocation substructures in both 1 at% Nb and 2 at% Al alloyed $MoSi_2$ predominantly exhibited the $1/2\langle 111]$ slip, with some $\langle 100]$ slip [19].

In a separate investigation, Misra et al. have observed an increase in the separation distance between $1/4\langle 111]$ partials by alloying with either Al or Nb [18]. Addition of 1 at% Nb to $MoSi_2$ increased the stacking width to 7.5 -8.5 nm for the 60° from screw orientation $1/4\langle 111]$ compared to $\approx 6.8$ for the unalloyed $MoSi_2$ reported by Ito et al. [17]. Furthermore, both Al and Nb solute atoms presumably segregate to the partials increasing the spacing between the partials and resulting in a lower stacking fault energy and hence, an enhanced dislocation mobility and a decreased flow stress.

Constant Yield Stress Temperature Range

Monocrystalline silicides exhibit anomalous increase in the yield stress with increasing temperature at different temperature ranges depending on the crystal orientations. In $MoSi_2$, this anomalous yield stress behavior is most pronounced for orientations favorable for $\{011\}\langle 100]$ and $\{010\}\langle 100]$ slip systems and occurs at the temperature range of about 600-900°C [23]. Similar behaviors are observed at temperature ranges of about 800-1100°C and 1000-1200°C for $\{110\}\langle 111]$ and $\{013\}\langle 331]$ slip systems, respectively. Therefore, the anomalous yield stress temperature range (AYSTR) in monocrystalline $MoSi_2$ samples varies from 600-1200°C depending on the sample orientation during compression testing.

In polycrystalline unalloyed $MoSi_2$, the AYSTR could not be determined conclusively due to the lack of plasticity below 900°C, however, it clearly falls below 900°C as evident from Fig. 2 and Fig. 3. Anomalous yielding behavior is observed in all alloyed samples. Of interest to this discussion is the temperature range of 600-800°C, previously called the CYSTR, with nearly constant yield stress of 325±25 MPa for all alloyed samples. The CYSTR is the plateau between the lower temperature region where yield stress decreases with temperature and higher temperature region of the AYSTR. Due to a change in the slope of yield strength vs. temperature line from negative to positive, a plateau is expected at this temperature range. However, the lack of an effect due to alloying on the yield stress is not expected. Alloying with Re, Al, Nb, or their combinations did not affect the yield stress of MoSi$_2$ at the CYSTR.

The cause of this anomalous behavior may be elucidated by considering the operative slip systems at temperatures about the CYSTR in unalloyed and alloyed MoSi$_2$. For unalloyed MoSi$_2$, below 600°C the operative slip systems are {011}⟨100], {013}⟨331], and {110}⟨111] [23]. Since this temperature range is below AYSTR, the CRSS decreases for all slip systems at this temperature range. At the CYSTR, i.e. 600-800°C, {011}⟨100], {110}⟨111], and {010}⟨100] slip systems are active and the CRSS increases for {011}⟨100] and {010}⟨100] and it decreases for {110}⟨111] [23]. Above CYSTR, most slip systems are active depending on the crystal orientations. In polycrystalline MoSi$_2$, since all the available slip systems at a specific temperature range are active in some grains, the yield stress should be a weighted average of the yield stress for all the individual slip systems. This weighted average is about the yield stress at the intersection of the plots of yield stress vs. temperature for the aforementioned slip systems, i.e. 325±25 MPa, Fig. 2 and Fig. 3. Alloying polycrystalline MoSi$_2$ with Al or Nb does not result in new slip systems [18,24] but shifts the temperature for activation of the 1/2⟨111] and ⟨100] type slips to lower temperatures, Fig. 4. As discussed in the previous section, below and above the CYSTR the addition of Al and Nb to MoSi$_2$ promote plasticity and hardening, respectively. At the CYSTR, the two competing phenomena are taking place simultaneously. As a result, at the CYSTR, these two competing phenomena nullify the effects of each other; hence, no significant change in the yield stress is observed at 600-800°C due to alloying.

SUMMARY

Alloying polycrystalline MoSi$_2$ with Re, Al, and Nb within the solubility limits in C11$_b$ structure had a significant effect on the yield strength and BDTT in compression. 2.5 at% Re increased the yield strength of MoSi$_2$ by more than an order of magnitude above at 1600°C. 2 at% Al or 1 at% Nb alloying resulted in ambient temperature ductility and high temperature strengthening. At the temperature range of 600-800°C, the yield strength of unalloyed and alloyed MoSi$_2$ remained constant at 325±25 MPa.

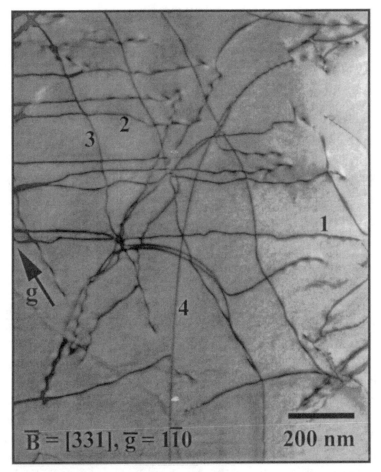

Fig. 4. Bright field TEM micrograph showing the dislocation substructure in a $MoSi_2$-1 at% Nb alloy compressed ≈0.5% at 400 °C. The labels 1 and 2 correspond to ⟨100] and 3 and 4 correspond to $1/2$⟨111] type dislocations, respectively.

REFERENCES

[1]K. Ito, T. Nakamoto, H. Inui, and M. Yamaguchi, "Stacking Faults on (001) in Transition-Metal Disilicides with the $C11_b$ Structure," pp. 599-604 in High-Temperature Ordered Intermetallic Alloys VII, Proceedings of the Materials Research Society Symposium, (Boston, MA, Dec. 2-5, 1996). Edited by C. C.

Koch, C. T. Liu, N. S. Stoloff, and A. Wanner Materials Research Society, Pittsburgh, PA, 1997.

[2]A. Misra, A. A. Sharif, J. J. Petrovic, and T. E. Mitchell, "Rapid Solution Hardening at Elevated Temperatures by Substitutional Re Alloying in $MoSi_2$," *Acta Mater.,* **48** [4] 925-32 (2000).

[3]K. Yanagihara, T. Maruyama, and K. Nagata, "Effect of Third Elements on the Pesting Suppression of Mo-Si-X Intermetallics (X = Al, Ta, Ti, Zr and Y)," *Intermetallics,* **4** [S1] S133-S9 (1996).

[4]A. Stergiou, P. Tsakiropoulos, and A. Brown, "The intermediate and high temperature oxidation behaviour of $Mo(Si_{1-x}Al_x)_2$ intermetallic alloys," *Intermetallics,* **5** [1] 69-81 (1997).

[5]S. A. Maloy, T. E. Mitchell, and A. H. Heuer, "High Temperature Plastic Anisotropy in $MoSi_2$ Single Crystals," *Acta Mater.,* **43** [2] 657-68 (1995).

[6]K. Ito, T. Yano, T. Nakamoto, H. Inui, and M. Yamaguchi, "Plastic Deformation of $MoSi_2$ and $WSi_2$ Single Crystals and Directionally Solidified $MoSi_2$-Based Alloys," *Intermetallics,* **4** [S1] S119-S131 (1996).

[7]U. V. Waghmare, E. Kaxiras, V. V. Bulatov, and M. S. Duesbery, "Effects of Alloying on the Ductility of $MoSi_2$ Single Crystals from First-Principles Calculations," *Model. Simul. Mater. Sci. Eng.,* **6** [4] 493-502 (1998).

[8]Y. Harada, Y. Muratam, and M. Morinaga, "Solid Solution Softening and Hardening in Alloyed $MoSi_2$," *Intermetallics,* **6** [6] 529-35 (1998).

[9]H. Inui, K. Ishikawa, and M. Yamaguchi, "Effects of Alloying Elements on Plastic Deformation of Single Crystals of $MoSi_2$," pp. 61-65 in Proceedings of the U.S.-Japan Workshop on Very High Temperature Structural Materials, (Turtle Bay, Hawaii, Dec. 9-11, 98). Edited by M. Yamaguchi and T. Pollock, 1998.

[10]P. Peralta, S. A. Maloy, F. Chu, J. J. Petrovic, and T. E. Mitchell, "Mechanical Properties of Monocrystalline $C11_b$ $MoSi_2$ with Small Aluminum Additions," *Scripta Mater.,* **37** [10] 1599-604 (1997).

[11]A. A. Sharif, A. Misra, J. J. Petrovic, and T. E. Mitchell, "Solid Solution Hardening and Softening in $MoSi_2$ Alloys," Accepted for publication *Scripta Mater.,* In Press (2001).

[12]Y. Harada, Y. Funato, M. Morinaga, A. Ito, and Y. Sugita, "Solid Solubilities of Ternary Elements and Their Effects On Microstructure of Mosi2," *J. Jap. Inst. Metal.,* **58** [11] 1239-47 (1994).

[13]S. Sastry, R. Suryanarayanan, and K. Jerina, "Consolidation and Mechanical Properties of $MoSi_2$-Based Materials," *Mater. Sci. Eng. A,* **192/193,** 881-90 (1995).

[14]R. Mitra and V.V.R. Rao, "Effect of minor alloying with Al on oxidation behaviour of MoSi2 at 1200 degrees C," *Mater. Sci. Eng. A,* **260** [1-2] 146-60 (1999).

[15]T. Nakano, M. Azuma, and Y. Umakoshi, "Microstructure and High-Temperature Strength in $MoSi_2$/$NbSi_2$ Duplex Silicides," *Intermetallics,* **6** [7-8] 715-22 (1998).

[16]H. Inui, M. Moriwaki, K. Ito, and M. Yamaguchi, "Plastic deformation of single crystals of Mo(Si ;Al)(2) with the C40 structure," *Phil. Mag. A,* **77** [2] 375-94 (1998).

[17]K. Ito, H. Inui, Y. Shirai, and M. Yamaguchi, "Plastic-Deformation of Mosi2 Single-Crystals," *Phil. Mag. A,* **72** [4] 1075-97 (1995).

[18]A. Misra, A. A. Sharif, J. J. Petrovic, and T. E. Mitchell, "Mechanical Behavior of Molybdenum Disilicide-Based Alloys," pp. In press Proceedings of the Materials Research Society Symposium, (Boston, MA, Nov27-Dec1, 2000). Edited by J.H. Schneibel, S. Hanada, K. J. Hemker, R. D. Noebe, G. Sauthoff Materials Research Society, Pittsburgh, PA, 2000.

[19]A.A. Sharif, A. Misra, J.J. Petrovic, and T.E. Mitchell, "Alloying of $MoSi_2$ for Improved Mechanical Properties," in Proceedings of the TMS Annual Meeting, (New Oreleans, LI, Feb 11-15) 2001.

[20]T. E. Mitchell and A. Misra, "Structure and Mechanical Properties of (Mo, Re) $Si_2$ Alloys," *Mater. Sci. Eng. A,* **261** [1-2] 106-12 (1999).

[21]J. W. Bray, *Aluminum Mill and Engineered Wrought Products,* pp. 29-61 in Metals Handbook V2 Properties and Selected Nonferrous Alloys, ASM International, New York, 1990.

[22]S.A. Maloy, A.H. Heuer, J.J. Lewanddowski, and T.E. Mitchell, "On the Slip Systems in $MoSi_2$," *Acta Metall. Mater.,* **40** [11] 3159-65 (1992).

[23]K. Ito, M. Moriwaki, T. Nakamoto, H. Inui, and M. Yamaguchi, "Plastic Deformation of Single Crystals of Transition Metal Disilicides," *Materials Science and Engineering A,* **233** [1-2] 33-43 (1997).

[24]A. A. Sharif, A. Misra, and T. E. Mitchell, "Effects of 1 at% Nb on Deformation Mechanisms of $MoSi_2$," *Manuscript to be published.*

# Reaction Forming

# IN SITU SYNTHESIS OF NONOXIDE-BORON NITRIDE (NOBN) COMPOSITES

G. J. Zhang*
Synergy Materials Research Center
National Institute of Advanced
Industrial Science and Technology
Nagoya 463-8687, Japan

J. F. Yang
Japan Science and Technology
Corporation (JST)
Nagoya 463-8687, Japan

T. Ohji and S. Kanzaki
Synergy Materials Research Center
National Institute of Advanced
Industrial Science and Technology
Nagoya 463-8687, Japan

## ABSTRACT

In situ process is advantageous for obtaining better ceramic composites. In this work the reactions and processing of in situ nonoxide-boron nitride (Nobn) composites include SiC-BN, $Si_3N_4$-BN, SiAlON-BN, AlN-BN and AlON-BN composites were discussed. For obtaining these BN composites, novel in situ reactions were proposed. Reaction mechanisms and the factors, which affect the reaction processes, were investigated and discussed according to the results of XRD and TG-MS analysis. It was pointed out that the furnace atmospheres and pressure demonstrated obvious effect on the reaction process. Finally, in situ Nobn composites were prepared by hot pressing or pressureless sintering according to the proposed in situ reactions. In some case net-shape sintering could be realized.

*Now with Synergy Ceramics Laboratory, Fine Ceramics Research Association (FCRA), Nagoya 463-8687, Japan.   Email: zh77@yahoo.com

# INTRODUCTION

For particulate ceramic composites, besides phase fraction, microstructural parameters of second phase particles, such as particle shape and size, distribution and homogeneity play an important role in the properties of the obtained composites and various processing approaches have been proposed. Particulate composites were mainly fabricated by sintering of mechanically-mixed powders of the component phases. In that process the distribution and homogeneity of the second phase particles in the matrix is strongly dependent on the shape and particle size of the starting powders and dispersing process. Another way to prepare particulate composites is in situ reaction synthesis, which is a powerful process to obtain better particulate composites with finer and more homogeneous microstructures, higher chemical and microstructural stability at high temperatures and better mechanical properties upon those obtained by conventional processes.[1, 2] Especially for composites with flake-shaped graphitic hexagonal BN second phase, because BN agglomerates or large BN flakes which are difficult to avoid their existence in starting powders, may act as fracture flaws due to the easy cleavage perpendicular to the c-axis of BN flakes[2-4], in situ formation of fine and homogeneously distributed BN flakes in ceramic matrix is an attractive way. Fig. 1(a) shows the micrograph of a BN powder, the existence of large BN flake can be clearly seen. Fig. 1(b) shows the microstructure of fracture surface of a SiC-BN composite prepared by hot pressing using the above BN powder. It can be seen that large BN flake exists in the material. It is suggested that the existence of large BN flakes in BN composites is one of the most important factors to decrease the material strength.

BN composites are well known as thermal shock and corrosion resistant materials, such as $Si_3N_4$-BN[4], Sialon-BN[3], AlN-BN[6], Alon-BN[7], SiC-BN[8] and $Al_2O_3$-BN[9]. Usually, with the increase of BN in the composites, the densification behavior became markedly poor and the strength of the composites decreased when they were prepared by conventional process. As mentioned above, in situ process is an effective way to obtain better BN composites. Coblenz and Lewis[9] obtained $SiO_2$-BN, $Al_2O_3$-BN and mullite-BN composites with homogeneous microstructures and good mechanical properties by the in situ reactions of $Si_3N_4 + B_2O_3$, $AlN + B_2O_3$ or $Si_3N_4 + AlN + B_2O_3$, respectively. Kusunose et al produced $Si_3N_4$-BN nanocomposites[10] by the reaction of $H_3BO_3$ and urea to synthesize nano-sized BN particles dispersed in $Si_3N_4$ matrix.

Innovative Processing and Synthesis of Ceramics

Fig. 1 Microstructures of (a) a BN raw powder (SP-2 grade, particle size 4 μm, Denki Kagaku Kogyo Co. Ltd., Japan) and (b) an SiC-25 vol% BN composite prepared with this BN powder by conventional process

Recently, we proposed some novel in situ reactions to prepare SiC-BN, Si₃N₄-BN, Si₃N₄-SiC-BN, AlN-BN and Alon-BN composites[11-15]. Composites with very fine microstructures and much higher strength could be fabricated by hot pressing or pressureless sintering the mixed reactant powders of the proposed in situ reactions.

## EXPERIMENTAL PROCEDURE

The starting powders were $Si_3N_4$ (E-10 grade, Ube Industries, Japan), $B_4C$ (Fl grade, particle size 1 μm, Denki Kagaku Kogyo Co. Ltd., Japan), C (2600# grade, Mitsubishi Chemical Co., Japan), $Al_2O_3$ (particle size 0.2 μm, Daimei Chemical Co., Japan), $Y_2O_3$ (Shin-etsu Chemical Co., Japan), α-SiC (A2 grade, mean particle size 0.63 μm, Showa Denko Co., Japan), h-BN (SP-2 grade, particle size 4 μm, Denki Kagaku Kogyo Co. Ltd., Japan), $SiB_6$ (particle size <5 μm, purity >99%, High Purity Chemicals Laboratory, Saitama, Japan) and $AlB_2$ (particle size <10 μm, purity >99%, High Purity Chemicals Laboratory, Saitama, Japan). For improving the sinterbility, $Al_2O_3$-$Y_2O_3$ or $Y_2O_3$ additives were used for different system.

For preparing SiC-BN and Si₃N₄-SiC-BN composites from reaction (1) and (3), hot pressing of the mixed powders was conducted in an argon or nitrogen atmosphere, respectively, under 30 MPa in a graphite die with BN coating. For other systems, pressureless sintering was conducted in nitrogen atmosphere. For studying the phase formation mechanism, hot pressing or pressureless sintering at different temperatures for 60 min was conducted and then the phase composition was determined by X-ray diffraction (XRD) using CuKα radiation. Thermogravimetry-differential thermal analysis (TG-DTA) was also conducted to study the reaction processes. The microstructures of the composites were observed by scanning electron microscopy (SEM).

## RESULTS AND DISCUSSION

Proposed in situ reactions

The proposed in situ reactions are as follows:

*SiC-BN system:*

$$Si_3N_4 + B_4C + 2C = 3SiC + 4BN \dots\dots\dots\dots\dots\dots\dots\dots\dots\dots\dots(1)$$

The weight percents of SiC and BN in the in situ SiC-BN composite according to reaction (1) are 54.79 % and 45.21 %, respectively, and the volume percents are 46.29 % and 53.71 %, respectively. The fraction of BN may be realized by adding SiC powder. Another in situ reaction may also be used to prepare SiC-BN composite:

$$SiB_6 + C + 3N_2 \rightarrow SiC + 6BN \qquad (2)$$

The weight percents of SiC and BN in the in situ SiC-BN composite according to reaction (2) are 21.21 % and 78.79 %, respectively, and the volume percents are 15.95 % and 84.05 %, respectively. The fraction of BN may also be realized by adding SiC powder.

*$Si_3N_4$-SiC-BN system:*

$$(1 + x)Si_3N_4 + B_4C + 2C = xSi_3N_4 + 3SiC + 4BN \qquad (3)$$

Namely, the BN content in the obtained composite can be controlled by adding extra-amount of $Si_3N_4$ in the starting mixed powders (*x* value).

*$Si_3N_4$-BN system:*

$$3SiB_6 + 11N_2 \rightarrow Si_3N_4 + 18BN \qquad (4)$$
$$3SiB_4 + 8N_2 \rightarrow Si_3N_4 + 12BN \qquad (5)$$

The weight percents of $Si_3N_4$ and BN in the in situ $Si_3N_4$-BN composite according to reaction (4) are 23.90 % and 76.10 %, respectively, and the volume percents are 18.29 % and 81.71 %, respectively. For reaction (5), the weight percents of $Si_3N_4$ and BN are 32.02% and 67.98%, respectively, and volume percents are 25.14% and 74.86%, respectively. The fraction of BN may also be realized by adding $Si_3N_4$ powder.

*AlN-BN system:*

$$AlB_2 + 1.5N_2 \rightarrow AlN + 2BN \qquad (6)$$

The weight percents of AlN and BN in the in situ AlN-BN composite according to reaction (6) are 45.23 % and 54.77 %, respectively, and the volume percents are 36.51 % and 63.49 %, respectively. The fraction of BN may also be realized by adding AlN powder.

In addition, by adding $Al_2O_3$ and/or $Si_3N_4$ to reactions (4), (5) and (6) Sialon-BN or Alon-BN can be prepared.

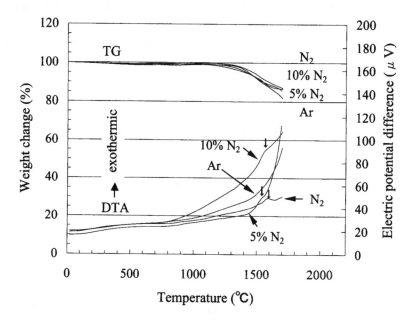

Fig. 2 TG-DTA results obtained from reaction (1) in argon with different partial pressure of nitrogen

Reaction process:

Some examples of the reaction process of the above reactions are discussed below. The TG-DTA results obtained from reaction (1) in argon with different partial pressure of nitrogen are showed in Fig. 2. In all atmospheres weight loss occurred from about 1150°C. However, the value of weight loss up to 1500°C decreased in nitrogen-containing atmosphere. It should mean that the existence of nitrogen had inhibited the decomposition of $Si_3N_4$ at high temperatures. On the other hand, there shows obvious difference in DTA curves from different atmospheres. Although with the limitation of analysis temperature by the apparatus, in atmospheres with low partial pressure of nitrogen a sharp exothermic peak can be deduced to exhibit at temperature higher than 1700°C. The exothermic reaction was remarkably inhibited in pure nitrogen atmosphere and before the large exothermic peak a small exothermic peak appeared at 1590°C. This small exothermic peak appeared to 1548°C and 1536°C in atmospheres with 10% and 5% nitrogen, respectively. In pure nitrogen atmosphere, only the small exothermic peak appeared. It is demonstrated that the rate of the exothermic reaction was very slow in pure nitrogen atmosphere. As shown in Fig. 3, however, the overall reaction (1) finished after hot pressing the mixed powders at 1800°C under 20 MPa for 60 min in nitrogen atmosphere.

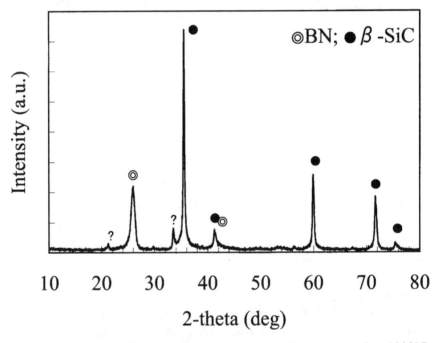

Fig. 3 XRD pattern of the product from reaction (1) hot pressed at 1800°C under 20 MPa for 60 min in nitrogen atmosphere

Innovative Processing and Synthesis of Ceramics

Fig. 4 shows the XRD patterns of the products of AlB$_2$ nitrided in nitrogen at different temperature, pressure and time. According to the TG-DTA result, the nitridation of AlB$_2$ took place very fast at about 1000°C and a very sharp exothermal peak appeared under ambient nitrogen atmosphere. Moreover, according to the peak intensity of XRD results showing in Fig. 4, in the nitridation of AlB$_2$ the formation of AlN was much faster than that of BN. With the pressure increase of N$_2$ atmosphere, the nitridation process was accelerated. It can be seen from Fig. 4 that under high pressure of N$_2$ atmosphere, only the main component phases of AlN and BN was detected by XRD.

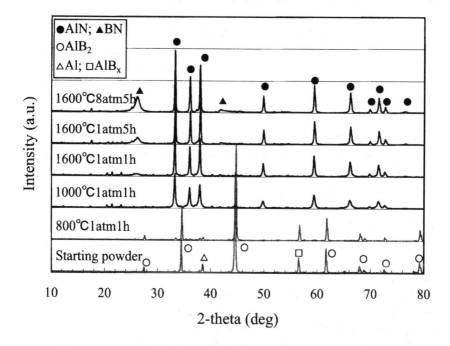

Fig. 4 XRD patterns of the products of AlB$_2$ nitrided in nitrogen at different temperature, pressure and time

However for the SiB$_6$, the nitridation process was slow. Although the nitridation of boron was fast, that of silicon was very slow and long period of time is needed for full nitridation of SiB$_6$.

In addition, due to the volume expansion of the nitridation of $SiB_6$ and $AlB_2$ are very large, during the nitridation process such volume expansion would partly compensate the shrinkage of the porous green body and consequently near-net shape sintering with low shrinkage and high density would be realized. For example, the leaner shrinkage of AlN-30 %BN and Alon-21 % BN was only about 3%.

## CONCLUSION

Various in situ reactions for preparing nonoxide-BN (Nobn) composites include SiC-BN, $Si_3N_4$-BN, SiAlON-BN, AlN-BN and AlON-BN composites were proposed and discussed. Reaction mechanisms and the factors, which affect the reaction processes, were investigated according to the results of XRD and TG-MS analysis. It was pointed out that the furnace atmospheres and pressure demonstrated obvious effect on the reaction process. Finally, in situ Nobn composites were prepared by hot pressing or pressureless sintering according to the proposed in situ reactions. In some case net-shape sintering could be realized.

### ACKNOWLEDGEMENT

This work has been supported by AIST, METI, Japan, as part of the Synergy Ceramics Project. The authors are members of the Joint Research Consortium of Synergy Ceramics.

### REFERENCES

[1]Y. M. Chiang, J. S. Haggerty, R. P. Messner and C. Demetry, "Reaction-Based Processing Methods for Ceramic-Matrix Composites," *Am. Ceram. Soc. Bull.,* **68** [2] 420-28 (1989).

[2]G. J. Zhang, Y. Beppu, T. Ohji and S. Kanzaki, "Reaction Mechanism and Microstructure Development of Strain Tolerant In Situ SiC-BN Composites," *Acta Mater.,* **49**, 77-82 (2001).

[3]W. Sinclair and H. Simmons, "Microstructure and Thermal Shock Behavior of BN composites," *J. Mater. Sci. Lett.,* 6, 627-29 (1987).

[4]D. Goeuriot-Launay, G. Brayet and F. Thevenot, "Boron Nitride Effect on the Thermal Shock Resistance of an Alumina-Based Ceramic Composite," *J. Mater. Sci. Lett.,* 5, 940-42 (1986).

[5]K.S.Mazdiyasni and R.Ruh, "High/Low Modulus $Si_3N_4$-BN Composite for Improved Electrical and Thermal Shock Behavior," *J. Am. Ceram. Soc.,*

**64** [7] 415-419 (1981).

[6]T. Kanai, K. Tanemoto and H. Kubo, "Hot-Pressed BN-AlN Ceramic Composites of High Thermal Conductivity," *Jpn. J. Appl. Phys.*, **29**, 683-87 (1990).

[7]A. Shimpo, M. Ueki and M. Naka, "Mechanical Properties of ALON-Based Composite Ceramics at High Temperature (Part 1)," *J. Ceram. Soc. Japan.*, **109**, 127-31 (2001).

[8]P. G. Valentine, A. N. Palazotto, R. Ruh and D. C. Larsen, "Thermal Shock Resistance of SiC-BN Composites," *Adv. Ceram. Mater.*, **1** [1] 81-87 (1986).

[9]W. S. Coblenz and D. Lewis, "In Situ Reaction of $B_2O_3$ with AlN and/or $Si_3N_4$ to Form BN-Toughened Composites," *J. Am. Ceram. Soc.*, **71** [12] 1080-85 (1988).

[10]T.Kusunose, Y.H.Chao, T.Sekino and K.Niihara, "Mechanical Properties of $Si_3N_4$/BN Composites by Chemical Processing," *Key Engineering Materials*, **161-163**, 475-480 (1999).

[11]G. J. Zhang and T. Ohji, "In Situ Reaction Synthesis of SiC-BN Composites," *J. Am. Ceram. Soc.*, **84** [7] (2001).

[12]G. J. Zhang and T. Ohji, "Effect of BN Content on Elastic Modulus and Bending Strength of SiC-BN In Situ Composites," *J. Mater. Res.*, **15** [9], 1876-80 (2000).

[13]G. J. Zhang, J. F. Yang, Z. Y. Deng and T. Ohji, "Effect of $Y_2O_3$-$Al_2O_3$ Additive on the Phase Formation and Densification Process of In Situ SiC-BN Composite," *J. Ceram. Soc. Japan.*, **109**, 45-48 (2001).

[14]G. J. Zhang, J. F. Yang, T. Ohji and S. Kanzaki, "In Situ Reaction Synthesis of Nonoxide-BN Composites," *Adv. Eng. Mater.*, in review (2001).

[15]G. J. Zhang, J. F. Yang and T. Ohji, "In Situ $Si_3N_4$-SiC-BN Composites: Preparation, Microstructures and Properties," *Mater. Sci. & Eng. A*, **2001**, accepted.

# FABRICATION OF LIGHTWEIGHT OXIDE/INTERMETALLIC COMPOSITES AT 1000°C BY THE DISPLACIVE COMPENSATION OF POROSITY (DCP) METHOD

Patrick J. Wurm, Pragati Kumar, Kevin D. Ralston, Michael J. Mills, and Ken H. Sandhage
Department of Materials Science and Engineering
481 Watts Hall, 2041 College Road
The Ohio State University
Columbus, OH 43210

## ABSTRACT

The Displacive Compensation of Porosity (DCP) method is a novel, reactive infiltration approach for fabricating dense, near net-shaped, ceramic/metal composites with high ceramic contents at modest temperatures. In this process, a metallic liquid is pressureless infiltrated into a porous, shaped ceramic preform and allowed to undergo a displacement reaction with the ceramic phase in the preform. Unlike other reactive infiltration approaches, a larger volume of ceramic is produced than is consumed, so that the prior pore volume within the preform becomes filled (i.e. reaction-induced densification). In this paper, the DCP method is used to fabricate MgO/Fe-Al intermetallic composites. Dense MgO/FeAl-based composites were produced by the pressureless infiltration of Mg(l) into, and then reaction with, porous $MgAl_2O_4$/MgO/Fe-bearing preforms within 2 h at 1000°C.

## INTRODUCTION

Intermetallic/ceramic composites can possess an attractive combination of properties. Consider, for example, composites of oxidation-resistant aluminide intermetallics (e.g., Ni-Al, Fe-Al) with lightweight refractory oxides. Such composites should possess lower weights, higher stiffnesses, and enhanced creep resistance relative to monolithic aluminides.[1-6] Such composites can also be much tougher than the monolithic oxides. Several authors have reported that aluminide/ceramic composites comprised of ≥65 vol% of the ceramic phase (e.g., Ni$_3$Al/Al$_2$O$_3$, FeAl/TiC composites) can possess fracture toughness values of 10-18 MPa·m$^{1/2}$.[6-9] Such high toughness values were reported to be the result of an increase in ductility of the intermetallic phase when this phase was present as thin, continuous filaments with a grain size larger than the filament thickness.[6,7]

Aluminide/oxide composites have been fabricated by powder metallurgical and melt infiltration approaches.[5-9] Dense composites produced by the sintering of shaped, powder-based preforms undergo shrinkage that results in a change of the dimensions and, if such shrinkage is not uniform, the shapes of the starting preforms. Infiltration of a metallic liquid into a rigid, porous preform, followed by solidification, can yield a dense composite that retains the preform shape and dimensions (i.e., near net-shape processing). However, melt infiltration processing of an oxidation-resistant aluminide, such as NiAl or FeAl, is complicated by the relatively high melting temperatures of the aluminide (e.g., T$_m$[NiAl] = 1638°C, T$_m$[FeAl] = 1330°C).[10,11] Over the past decade, several reactive infiltration or penetration approaches have been developed to overcome these difficulties. Such methods include: the Reactive Metal Penetration (RMP) process, the Co-Continuous Ceramic Composite (C4) process, and the infiltration Alumina Aluminide Alloy (i-3A) process.[12-15] In each of these approaches, a low-melting metallic liquid (e.g., Al(l)) undergoes a displacement (oxidation-reduction) reaction with a ceramic preform to yield new solid metal and oxide products. By tailoring the composition of the preform and/or the liquid (i.e., by using alloy liquids), dense, near net-shaped aluminide/oxide composites (e.g., composites of alumina with TiAl$_3$, NiAl, or NbAl$_3$) can be produced.[12,14,15] For the RMP, C4, and i-3A processes, however, the ceramic content of the product composite is less than the ceramic content of the preform (i.e., less ceramic is produced than is consumed by the displacement reaction). As mentioned above, prior work with FeAl-bearing and Ni$_3$Al-bearing composites has shown that a high ceramic content is desired in order to obtain a ductile, filamentary aluminide phase with a small filament thickness.[6,7]

A novel, reactive infiltration approach for fabricating dense, near net-shaped, ceramic-rich composites has recently been developed at The Ohio State

University.[16-19] This process, known as the Displacive Compensation of Porosity[1] (DCP) method, involves the use of displacement reactions that yield a larger volume of ceramic than is consumed. The purpose of this paper is to demonstrate that lightweight MgO/FeAl-based intermetallic composites can be produced at 1000°C by the DCP process. Such composites should exhibit enhanced oxidation and creep resistance, and reduced weight, relative to monolithic FeAl (since $\rho[MgO]$ = 3.58 g/cm$^3$, and $\rho[FeAl]$ = 6.05 g/cm$^3$).[1] Prior work has also shown that dense, MgO-rich composites produced by the DCP method are quite resistant to hydration.[17]

EXPERIMENTAL PROCEDURE

Composites of MgO and FeAl were fabricated by the pressureless infiltration of Mg liquid into, and reaction with, porous MgAl$_2$O$_4$/MgO/Fe preforms.

Fe powder (ave. size = 1-3 $\mu$m, >98% purity) was milled for 4 hours in a high-energy vibratory ball mill (Model 8000 Mixer/Mill, SPEX Industries, Edison, NJ), using 20 ml of a distilled hexane solution containing 0.1 wt% stearic acid (95% purity, Aldrich Chemical Company, Inc., Milwaukee, WI) as a milling lubricant. The MgAl$_2$O$_4$ (ave. size $\leq$ 44 $\mu$m, 99.9% purity, Cerac, Inc., Milwaukee, WI) and MgO powders (ave. size $\leq$ 44 $\mu$m, 99.95% purity, Johnson-Matthey, Inc., Ward Hill, MA) were then added to achieve a MgAl$_2$O$_4$:MgO:Fe molar ratio of 1.0:2.0:3.0. This powder mixture was then milled for an additional 2.5 hours. Bar-shaped (0.2 cm x 3.0 cm x 0.7 cm) MgAl$_2$O$_4$/MgO/Fe preforms were produced by uniaxial pressing of this powder mixture at a peak stress of 2.1 MPa. After removing the hexane/stearic acid solution by heating for 4 h at 500°C under flowing Ar, the preforms were fired for 2 h at 1200°C under flowing Ar. After this heat treatment, the preforms were rigid and possessed bulk densities of 3.50±0.30 g/cm$^3$ (i.e., 74.9±6.4% of the theoretical density).[1]

For the infiltration experiments, bars of magnesium (99.9% purity, Johnson-Matthey, Inc.) were placed above and below the porous, bar-shaped MgAl$_2$O$_4$/MgO/Fe preforms within a magnesia crucible (2.5 cm wide x 10 cm long x 1.3 cm high, Ozark Technical Ceramics, Inc., Webb City, MO). The

---

[1]Also known as the PRIMA-DCP (Pressureless Reversible Infiltration of Molten Alloys by the Displacive Compensation of Porosity) method.[19]

specimen-bearing crucible was then wrapped and sealed within steel foil to minimize the loss of magnesium by vaporization during subsequent heating. The sealed assembly was heated at 8°C/min to 1000°C for 2 h under a flowing Ar atmosphere. The assembly was then cooled at a rate of 8°C/min to room temperature under the flowing Ar atmosphere. The microstructure and microchemistry of the resulting composites were characterized with a field emission gun scanning electron microscope (Model XL-30 FEG-SEM, Philips Electronics N.V., Eindhoven, The Netherlands) equipped with a Si/Li detector (Edax International, Mahwah, NJ) for energy-dispersive x-ray spectroscopy (EDS). The phase content was also evaluated by x-ray diffraction (XRD) analysis using Cu-K$\alpha$ radiation with a scan rate of 1°/min.

RESULTS & DISCUSSION

The following net reaction between Mg liquid and a $MgAl_2O_4$/MgO/Fe preform is thermodynamically favored at 1000°C:

$$3Mg(l) + [MgAl_2O_4(s) + 2MgO(s) + 3Fe(s)] = 6MgO(s) + 5Fe_{0.60}Al_{0.40}(s).$$

The brackets [ ] refer to the solid phases present within a porous preform. Hence, upon infiltration, the Mg(l) can undergo a displacement reaction with $MgAl_2O_4(s)$ to yield MgO(s) and Al(l). The liberated aluminum can then react with solid iron in the preform to produce a $Fe_{0.60}Al_{0.40}$ intermetallic compound. The volume change associated with the net conversion of a mixture of 1 mole of $MgAl_2O_4$, 2 moles of MgO, and 3 moles of Fe into a mixture of 6 moles of MgO and 5 moles of $Fe_{0.60}Al_{0.40}$ is +24.6%.[1] This reaction-induced increase in solid volume can be used to fill the pores within a rigid $MgAl_2O_4$/MgO/Fe preform. As this reaction proceeds, residual excess Mg(l) should gradually be extruded out of the rigid preform ("reversible infiltration"), until a dense MgO/$Fe_{0.60}Al_{0.40}$-based composite is produced.

Secondary and backscattered electron images of a MgO/$Fe_{0.60}Al_{0.40}$-based composite produced at 1000°C are shown in Figs. 1 and 2, respectively. The secondary electron image reveals a relatively dense microstructure with a small amount of closed, fine (<1 μm dia.) porosity. The backscattered electron image

Innovative Processing and Synthesis of Ceramics

reveals two distinct phases dispersed uniformly throughout the composite. EDS analyses indicated that the dark and light phases were MgO and FeAl, respectively. These phases were also detected by XRD analyses. The average composition obtained for the FeAl phase was $61.7\pm0.5$ at% Fe, $38.3\pm0.5$ at% Al, which was close to the composition expected from the stoichiometry of the reaction above. The FeAl phase possessed a flake-shaped morphology with a flake thickness of $\leq2$ μm (i.e., below the thickness for a brittle-to-ductile transition in FeAl reported by Subramanian and Schneibel for tough TiC/FeAl composites).[6] Quantitative image analysis of backscattered electron micrographs indicated that the composite consisted of $58.9\pm1.3$ vol% MgO, $38.8\pm1.9$ vol% FeAl, and $2.3\pm0.8$ vol% Mg. In order to further reduce the small amount of residual magnesium (not evident in the secondary and backscattered electron images in Figs. 1 and 2) detected in these specimens, subsequent reactive infiltration experiments were conducted on preforms with slightly higher relative densities. The XRD pattern obtained from a composite derived from a 82.5% dense preform is shown in Fig. 3. Unlike composites produced from more porous preforms, this composite did not exhibit diffraction peaks for magnesium; that is, the increase in internal solid volume associated with the displacement reaction was sufficient in this latter case to remove essentially all of the molten magnesium from this relatively dense specimen.

The reactive conversion of porous $MgAl_2O_4$/MgO/Fe preforms into dense MgO/$Fe_{0.60}Al_{0.40}$-based composites was completed within 2 hours at 1000°C. Shorter reaction times have not yet been examined. Nonetheless, this reaction temperature is about 400°C lower than that required to melt and infiltrate a $Fe_{0.60}Al_{0.40}$ composition into porous MgO preforms (the liquidus temperature for $Fe_{0.60}Al_{0.40}$ is about 1408°C).[10] Pressures in excess of 1 atm were also not required for melt infiltration with the present DCP process.

Prior work with the DCP method has shown that dense, MgO-rich composites can be produced that retain the shapes and dimensions of the preforms.[16-18] Work is currently underway to determine whether complex and near net-shaped, MgO/$Fe_{0.60}Al_{0.40}$-based composites can also be produced by the DCP process. The fracture strength and fracture toughness values of these composites are also being evaluated.

## CONCLUSIONS

MgO/$Fe_{0.60}Al_{0.40}$-based composites have been fabricated within 2 h at 1000°C by the DCP process. Liquid magnesium was pressureless infiltrated into porous, bar-shaped $MgAl_2O_4$/MgO/Fe preforms. Upon infiltration, the Mg(l) underwent a displacement reaction with $MgAl_2O_4$(s) to yield MgO(s) and Al(l). The liberated aluminum reacted with solid iron in the preform to produce an $Fe_{0.60}Al_{0.40}$ intermetallic compound. The increase in solid volume associated with this net reaction was compensated by the prior pore volume within the rigid preform. The resulting co-continuous composites were dense and exhibited a uniform distribution of MgO and a thin ($\leq 2$ μm), flake-shaped $Fe_{0.60}Al_{0.40}$ intermetallic phase.

## ACKNOWLEDGMENTS

The technical assistance of Mr. Cameron Begg (SEM analyses), Mr. Henk Colijn (XRD analyses), Mr. Steve Bright (image analyses), Mr. Gary Dodge (thermal treatments), and Mr. Ken Kushner (specimen grinding/machining) is greatly appreciated. The encouragement from, and technical discussions with, Ms. Susahn Briggs, Mr. Matt Dickerson, and Dr. Pragati Kumar are also gratefully acknowledged. Financial support provided by the U.S. Department of Energy through Grant # DE-FG02-96ER45550 (Dr. Yok Chen, Program Manager) is acknowledged.

## REFERENCES

[1]Powder Diffraction File Card Nos. 45-946 for MgO, 21-1152 for $MgAl_2O_4$, 6-696 for Fe, and 45-982 for FeAl. International Centre on Diffraction Data, Newtown Square, PA.

[2]J. H. Hensler, G. V. Cullen, "Stress, Temperature, and Strain Rate in Creep of Magnesium Oxide," *J. Am. Ceram. Soc.*, **51** [10] 557-559 (1968).

[3]G. R. Terwilliger, H. K. Bowen, R. S. Gordan, "Creep of Polycrystalline MgO and MgO-$Fe_2O_3$ Solid Solutions at High Temperatures," *J. Am. Ceram. Soc.*, **53** [5] 241-251 (1970).

[4]N.S. Stoloff, "Iron Aluminides: Present Status and Future Prospects," *Mater. Sci. Eng.* A, **A258**, 1-14 (1998).

[5]J.H. Schneibel, P.F. Becher, "Iron and Nickel Aluminide Composites," *J. Chinese Institute of Engineers*, **22**, 1-12 (1999).

[6]R. Subramanian, J. H. Schneibel, "The Ductile-Brittle Size Transition of Iron Aluminide Ligaments in an FeAl/TiC Composite," *Acta. Mater.*, **46**, 4733-4741 (1998).

[7]J. Rodel, H. Prielipp, N. Claussen, M. Sternitzke, K. B. Alexander, P. F. Becher, J. H. Schneibel, "$Ni_3Al/Al_2O_3$ Composites with Interpenetrating Networks," *Scripta Met. Mater.*, **33** [5] 843-848 (1995).

[8]K. P. Plucknett, P. F. Becher, R. Subramanian, "Melt-Infiltration Processing of $TiC/Ni_3Al$ Composites," *J. Mater. Res.*, **12** [10] 2525-2517 (1997).

[9]P. F. Becher, K. P. Plucknett, "Properties of $Ni_3Al$-bonded Titanium Carbide Ceramics," *J. Euro. Ceram. Soc.*, **18**, 395-400 (1997).

[10]*Binary Alloy Phase Diagrams*, 2$^{nd}$ Edition, Ed. T. B. Massalski, ASM International, Materials Park, OH, 1990.

[11]P. Nash, M. F. Singleton, J. L. Murray, in *Phase Diagrams of Binary Nickel Alloys*, Ed. P. Nash, ASM International, Materials Park, OH, 1991.

[12]W. G. Fahrenholtz, K. G. Ewsuk, R. E. Loehman, A. P. Tomsia, "Formation of Structural Intermetallics by Reactive Metal Penetration of Ti and Ni Oxides and Aluminates," *Met. Mater. Trans.*, **27A** [8] 2100-2104 (1996).

[13]M. C. Breslin, J. Ringnalda, L. Xu, M. Fuller, J. Seeger, G. S. Daehn, T. Otani, H. L. Fraser, "Processing, Microstructure, and Properties of Co-Continuous Alumina-Aluminum Composites," *Mater. Sci. Eng. A*, **A195**, 113-119 (1995).

[14]F. Wagner, D.E. Garcia, A. Krupp, N. Claussen, "Interpenetrating $Al_2O_3$-$TiAl_3$ Alloys Produced by Reactive Infiltration," *J. Euro. Ceram. Soc.*, **19**, 2449-53 (1999).

[15]C. Scheu, G. Dehm, W. D. Kaplan, F. Wagner, N. Claussen, "Microstructure and Phase Evolution of Niobium-Aluminide-Alumina Composites Prepared by Melt Infiltration," *Phys. Stat. Sol.(A)*, **166**, 241-255 (1998).

[16]P. Kumar, K. H. Sandhage, *U.S. Patent Application*, under review.

[17]P. Kumar, K. H. Sandhage, "The Displacive Compensation of Porosity (DCP) Method for Fabricating Dense, Shaped High-Ceramic-Bearing Bodies at Modest Temperatures," *J. Mater. Sci.*, **34**, 5757-5769 (1999).

[18]K. A. Rogers, P. Kumar, R. Citak, K. H. Sandhage, "The Fabrication of Dense, Shaped Ceramic/Metal Composites at ≤1000°C by the Displacive Compensation of Porosity Method," *J. Am. Ceram. Soc.*, **82**[3] 757-60 (1999).

[19]M. B. Dickerson, R. R. Unocic, K. T. Guerra, M. J. Timberlake, K. H. Sandhage, "The Fabrication of Dense Carbide/Refractory Metal Composites of Near Net Shape at Modest Temperatures by the PRIMA-DCP Process," pp. 25-31 in *Innovative Processing and Synthesis of Ceramics, Glasses, and Composites IV*, Ceram. Trans., Vol. 115, The American Ceramic Society, Westerville, OH 2000.

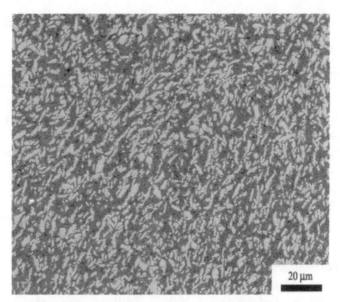

Figure 1. Secondary electron image of a polished cross-section of a MgO/FeAl composite formed at 1000°C within 2 hours.

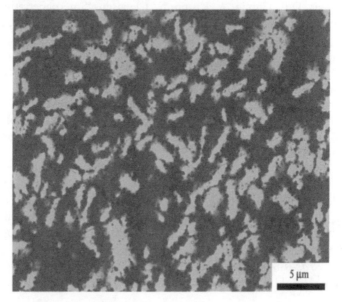

Figure 2. Backscattered electron image of a polished cross-section of a MgO/FeAl composite formed at 1000°C within 2 hours.

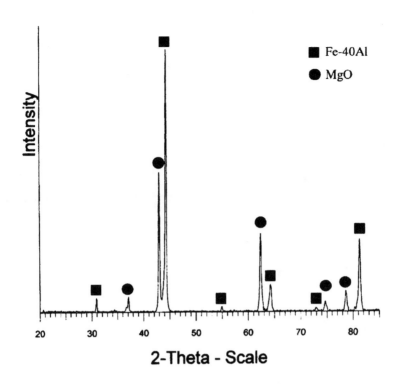

Figure 3. XRD pattern from a MgO/ $Fe_{0.60}Al_{0.40}$-based composite derived from a 82.5% dense preform.

# MICROSTRUCTURE AND MECHANICAL PROPERTIES OF SiAlON CERAMICS WITH POROUS STRUCTURE

Jianfeng Yang
Japan Science and Technology Corporation (JST), Nagoya 463-8687, Japan.
Yoshihisa Beppu
Synergy Ceramics Laboratory, Fine Ceramics Research Association, Nagoya 463-8687, Japan.
Guo-Jun Zhang and Tatsuki Ohji
Synergy Materials Research Center, National Institute of Advanced Industrial Science and Technology (AIST), Nagoya 463-8687, Japan.

## ABSTRACT

SiAlON ($Si_{6-z}Al_zO_zN_{8-z}$ ($z$=0.5-4.0)) ceramics with porous structure was produced using powder mixtures containing alpha $Si_3N_4$, AlN and $Al_2O_3$. With increasing in the z value, relative density was increased, however, further increasing the z to 4 resulted in a decreasing in relative density. The relative density ranged from 50% to 75%. X-ray analysis indicated that the only SiAlON phase was resulted. In this study we also investigated the corrosion behaviors of the porous and dense bodies of several types of SiAlONs in hydrochloric acid solutions. Changes of mass and strength after leaching was measured and the corrosion resistance was evaluated.

## INTRODUCTION

Porous ceramics have been expected to be used in many applications at high temperature, such as fluid-flow filters, lightweight structural components, electrodes, bioreactors, catalysts, etc., due to their superior properties, such as high-temperature stability, erosion/corrosion resistance.[1,2] Ceramic bulk filters with excellent high-temperature/chemical stability are strongly demanded for new power-generation systems, e.g., pressurized fluidized bed combustion (PFBC) and the integrated coal gasification combined cycle (IGCC).[3]

In the PFBC system, combustion takes place at temperatures from 800-900°C resulting in reduced NOx formation. The $SO_2$ emissions can be reduced by the injection of sorbent into the bed, and the subsequent removal of ash together with reacted sorbent. Coordinate or dolomite are commonly used for this purpose. However, problems arose that the thermal shock resistance is not so good for the reverse washing is difficult. Due to high temperature and corrosion environment, good corrosion resistance is strongly demanded.

The β-SiAlON ceramics, $Si_{6-z}Al_zO_zN_{8-z}$, is an alloy of silicon nitride and aluminum oxide. This superior refractory material has the combined properties of silicon nitride (high strength, hardness, fracture toughness and low thermal expansion) and aluminum oxide (corrosion resistance, chemically inert, high temperature capabilities and oxidation resistance).[4,5,6] Thus this kind of ceramics should be a good candidate for the high temperature filter.

In this study, we have attempted to synthesize the pure β-SiAlON ceramics with porous structure of three-dimensional (3-D) network via pressureless sintering. The SiAlON ceramics with z=0.5, 1, 2, 3 and 4 was obtained from the powder mixture of α-$Si_3N_4$, AlN and $Al_2O_3$ and sintered at 1750, 1800, and 1850°C. Reaction behavior, microstructure, and mechanical properties of the porous ceramics are reported.

## EXPERIMENTAL PROCEDURE

Five SiAlON materials with different stoichiometric compositions (SiAlON; z = 0.5, 1, 2, 3, 4) were prepared by pressureless sintering of highly pure submicrometer-sized powder mixture of $Si_3N_4$ (SN-E10, UBE Industries, Ltd., Tokyo, Japan; α ratio >95.5%, 1.3 wt% oxygen, mean particle size 0.55 μm), AlN (F-Grade, Tokuyama Corp., Tokyo, Japan, mean particle size 1.8 μm) and $Al_2O_3$ (TMD, Taimei Chemicals Co., Nagano, Japan). The composition of the starting powder is shown in Table I.

The mixture was wet-milled in methanol for 24 h. After dried and sieved to 150 mesh, the resulting powder mixtures were uniaxially pressed at 2 MPa, into rectangular plates measuring 55×35×10mm$^3$. The pressed green compacts were placed in a BN case and sintered in a graphite resistance furnace (Model No. FVPHP-R-5, Fujidempa Kogyo Co., Ltd., Osaka, Japan) with a powder bed of

Si$_3$N$_4$-SN powder mixture at a temperatures of 1800°C. The heating rate was 10-20°C/min, holding times were 2 h, and the nitrogen gas pressure of 0.6 MPa.

The bulk density of the sintered materials was measured by the Archimedes displacement method. Samples were measured in distilled water. The porosity was calculated from the ratio of the measured density and theoretical density, which were determined by rule of mixtures. The specimens were cut and machined into 40-50×4×3 mm$^3$ test bars, for flexural-strength measurement according to the JIS R1601 (Japanese Industrial Standard). All surfaces were finally ground on an 800-grit diamond wheel. After beveled, the specimens were loaded to failure in a three-point-bend mode (with a span of 30 mm) on a testing machine (Autograph AG-10TC, Shimadzu Corp., Kyoto, Japan.), at a cross-head speed of 0.5 mm/min.

Table 1. Composition of the SiAlON ceramics

| z | Si$_3$N$_4$ (wt%) | AlN (wt%) | Al$_2$O$_3$ (wt%) |
|-----|------|------|------|
| 0.5 | 91.5 | 2.4 | 6.1 |
| 1.0 | 83.0 | 4.9 | 12.1 |
| 2.0 | 66.2 | 9.7 | 24.1 |
| 3.0 | 49.5 | 13.5 | 36.0 |
| 4.0 | 32.9 | 19.2 | 47.9 |

X-ray diffraction (XRD) analysis of the specimens was performed after the machining, using CuK$_\alpha$ radiation (XRD; Model RINT2000, Rigaku Co., Ltd., Tokyo, Japan.) at 40 kV and 100 mA. The microstructure was characterized by scanning electron microscopy (SEM; Model No. S-5000, Hitachi, Ltd., Tokyo, Japan.). Fracture surfaces of the bending test were used for the observation after coated with Au.

Twelve specimens were soaked into 1 $M$ HCl solution in a polytetrafluoroethylen container of 150 mL (12.5 mL/sample). The container was then placed in a dry oven, and was kept at 70°C for 120 hours. After leaching, the specimens were taken from the container and were washed in water and dried. The three-point flexural strength measurements were carried out at room temperature before and after leaching, according to the JIS R1601. Three to seven specimens were used for each measurement. Changes of the mass and porosity during leaching were also determined. The porosity was measured by the Archimedes displacement method, using distilled water. The microstructure was characterized by scanning electron microscopy (SEM).

**RESULTS AND DISCUSSION**

Density measurement, XRD data and microstructural characteristics of the SiAlON materials after sintering at 1800°C are given in Table II as a function of the nominal z value. The density increased with the z value, and the highest density was obtained for the z = 3 material. All the five SiAlON compositions investigated were composed of the hexagonal'-phase only. The higher the z value the larger the grain size, as estimated by image analysis, which is similar with the dense SiAlON ceramics.[7]

It was known that SiAlON materials are easier to densify and more ductile at high temperatures than $Si_3N_4$.[6,8] The formation of the porous microstructure in the SiAlON ceramics in this study can be explained by the solution of the three phase of $Si_3N_4$, AlN and $Al_2O_3$ to form the single crystal of $Si_{6-z}Al_zO_zN_{8-z}$. After the formation of the SiAlON phase, no other oxide can be remained due to the mixture with stoichiometric compositions and no other sintering additives used. Thus further densification is retarded and porous microstructure formed.

Table II: Porous SiAlON ($Si_{6-z}Al_zO_zN_{8-z}$) characteristics

| z | Density (TD%) | Crystallized phase | Grain size (μm) |
|---|---|---|---|
| 0.5 | 50.4 | $Si_{0.5}Al_{5.5}O_{7.5}N_{0.5}$ | 0.5 |
| 1.0 | 59.1 | $SiAl_5O_7N$ | 1.0 |
| 2.0 | 60.3 | $Si_2Al_4O_6N_2$ | 1.5 |
| 3.0 | 75.4 | $Si_3Al_3O_5N_3$ | 2.0 |
| 4.0 | 63.5 | $Si_4Al_2O_4N_4$ | 3.0 |

SEM micrographs of the fracture surface are shown in Figs. 1(a), (b), for SiAlON materials with nominal z values of 1 and 3, respectively. The sintered porous SiAlON ceramics showed an increase in grain-size with increasing the nominal z value, as listed in Table I, while they were typically equiaxed

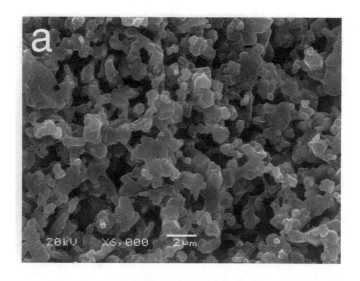

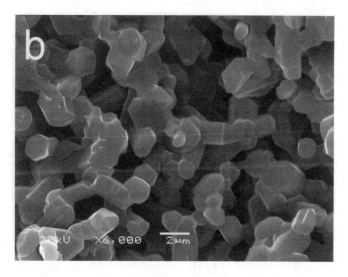

Figure 1.    SEM micrographs of the $Si_{6-z}Al_zO_zN_{8-z}$ porous ceramics with z = 1, 3 are shown in (a) and (b), respectively.

irrespective of the z value. As an exception, a very small fraction of elongated grains was recognized in the z = 1 SiAlON material. The average grain size of this material was 1.0 um; such a grain size was approximately twice of that measured in a high-purity $Si_3N_4$ material (i.e., 1 m for z = 0) also sintered with additive of $Y_2O_3$ and $Al_2O_3$ at 1800°C.[9] The z = 3 SiAlON reached a grain size after sintering about one order of magnitude higher than that of the material with z = 1, proving a significantly lowered-bulk refractoriness of the grains with increasing the z value. From the SEM micrographs, no other phase precipitated during the sintering can be observed, which was in agreement with the XRD results.

Table III. Properties of the porous (P) and dense (D) SiAlON samples before and after the corrosion by HCl solution

| z | Powder mixture ratio | Porosity (%) | Strength (MPa) | Mass loss ratio (%) | Strength loss ratio (%) |
|---|---|---|---|---|---|
| 0.5 (P) | 6.1wt% $Al_2O_3$ + 2.4wt% AlN | 48.7 | 30 (2) | 99.4 | 100 |
| 1.0 (P) | 12.1wt% $Al_2O_3$ + 4.9wt% AlN | 40.8 | 94 (19) | 99.4 | 98.1 |
| 2.0 (P) | 24.1wt% $Al_2O_3$ + 9.7wt% AlN | 34.2 | 97 (28) | 98.7 | 99.2 |
| 0.5 (D) | 6.1wt% $Al_2O_3$ + 2.4wt% AlN | 0.4 | 1046 (78) | 99.0 | 91.4 |
| 1.0 (D) | 12.1wt% $Al_2O_3$ + 4.9wt% AlN | 0.3 | 429 (160) | 100 | 100 |

The number in parentheses is the standard deviation.

The relative density and flexure strength of the as-sintered specimens (before leaching), and loss ratios of the flexural strength and mass of the specimens after leaching are listed in Table III. The ratio was determined by dividing the value after leaching by the original one. It showed that both the mass loss and the porosity increase were within 1%. All the strength was almost not degraded by leaching. Comparing with the porous ceramics with controlled sintering additive, in which the mass loss were about 5% and a great strength loss had been demonstrated[10], both the porous and dense SiAlON ceramics demonstrated excellent corrosion resistance. Such a small deterioration of the properties suggests that the SiAlON specimens have strong grain boundaries. In actuality, it has been observed in a transmission electron microscopy that porous SiAlON materials fabricated by the same procedures as this study have clean grain boundaries with little glassy phase.[11] The excellent corrosion resistance against the acidic solution indicates that the porous SiAlON can be a good candidate for filter materials usable in severe corrosive environments.

## SUMMARY

Fabrication, microstructures and mechanical properties of porous β-SiAlON ceramics have been investigated in five high-purity SiAlON ceramics with nominal z = 0.5, 1, 2, 3 and 4. With increasing in the z value, relative density was increased, however, further increasing the z to 4 resulted in a decreasing in relative density. The relative density ranged from 50% to 75%. The SiAlON materials mainly consisted of only hexagonal'-phase and no second glass phase was found by XRD and SEM investigation. The higher z value resulted in the significant grain growth. In this study we also investigated the corrosion behaviors of the porous and dense bodies of several types of SiAlONs in hydrochloric acid solutions. Both the porous and dense SiAlON materials demonstrated excellent corrosion resistance. Even for the porous bodies, there were no appreciable changes of mass and strength. It was suggested that the excellent corrosion resistance was due to the strong grain boundaries of the materials.

## ACKNOWLEDGEMENT

This work has been supported by METI, Japan, as part of the Synergy Ceramics Project. Part of the work has been supported by NEDO. The authors are members of the Joint Research Consortium of Synergy Ceramics.

## REFERENCES

[1] R. W. Rice, *Porosity of Ceramics*, Marcel Dekker, Inc., New York, 1998.

[2] K. Ishizaki, S. Komarneni, and M. Nanko, *Porous Materials*, (Kluwer Academic Publishers, Dordrecht, The Netherlands, 1998).

[3] K. Schulzk and M. Durst, "Advanced and an Integrated System for Hot Gas Filtration Using Rigid Ceramic Elements," *Filtration & Separation*, 31[1] 25-28 (1994).

[4] K. H. Jack, "Review: Sialons and Related Nitrogen Ceramics," *J. Mater. Sci.*, 11, 1135-58 (1976).

[5] K. H. Jack and W. I. Wilson, "Ceramics Based on the Si-Al-O-N and Related Systems," *Nature* (London) Phys. Sci., 283 [7] 28-29 (1972).

[6] T. Ekström and M. Nygren, "SiAlON Ceramics," *J. Am. Ceram. Soc.*, 75 [2] 259-76 (1992).

[7] G. Pezzotti, H.-J. Kleebe, K. Okamoto, and K. Ota, "Structure and Viscosity of Grain Boundary in High-Purity SiAlON Ceramics," *J. Am. Ceram. Soc.*, 83 [10] 2549-55 (2000)

[8] I. Tanaka, S. Nasu, H. Adachi, Y. Miyamoto, and K. Niihara, "Electronic Structure behind the Mechanical Properties of β-SiAlONS," *Acta Metall. Mater.*, 40 [8] 1995-2001 (1992).

[9] J.-F. Yang, T. Ohji and K. Niihara, "Influence of Yttria – Alumina Content on Sintering Behavior and Microstructure of Silicon Nitride Ceramics," *J. Am. Ceram. Soc.*, 83 [8] 2094-96 (2000).

[10] Y. Beppu, J.-F. Yang and T. Ohji, "Corrosion Resistance of Porous Silicon Nitride and SiAlON Ceramics," *Ceram. Tran.*, in press.

[11] J.-F. Yang and T. Ohji, "Fabrication and Properties of Porous SiAlON Ceramics," unpublished work.

# Functionally Graded Materials & Coatings

# FUNCTIONALLY GRADED POROSITY IN CERAMICS - ANALYSIS WITH HIGH RESOLUTION COMPUTED TOMOGRAPHY

J. Goebbels and G. Weidemann
Bundesanstalt fuer Materialforschung
und -pruefung (BAM)
Unter den Eichen 87
12205 Berlin, Germany

R. Dittrich, M. Mangler and
G. Tomandl
TU Bergakademie Freiberg
Gustav-Zeunerstrasse 3
09596 Freiberg, Germany

## ABSTRACT

The characterization of graded materials requires methods mapping local varying properties of materials. One of the most useful tools is high-resolution computed tomography. The ability to view inside the samples could be used to determine several features like density variation, porosity distribution and other structural parameters. With one of the two 3D-tomographs developed at BAM, a spatial resolution of about 2 µm can be reached dependent of the object diameter. As radiation source, a micro focus X-ray tube with a transmission target is used together with a cooled CCD camera and a fiber optic coated with a thin scintillate layer.

Ceramic membranes with capillary pores (e.g. Hydroxylapatite, $TiO_2$) are analyzed with CT and image processing. Beside the variation of pore diameter, which are in the range of 3 to 30 µm, the throughput of the capillaries and their structure was measured and analyzed.

## INTRODUCTION

Porous ceramics have been widely used as industrial filters, catalyst supports and gas or chemical sensors. The aim of this project was to find out the basic procedure for preparing porous ceramics with uniform capillaries by a process using the alginate gelation method. This process allows the production of structured $Al_2O_3$, $TiO_2$, $ZrO_2$ and hydroxylapatite ceramics.

In this study, alginate was mixed with ceramic slurry. Alginate is well known as an organic polymer, which can be gelatinated by cross-linking with multivalent metal ions. After this gelation the gel is dried and sintered. Porous ceramics with a unique microstructure of uniform capillaries and narrow pore size distributions were obtained.

To characterize functionally graded materials, methods are desirable which give a three-dimensional image of local properties like inhomogenities, volume distribution of elements, pores and cracks. Due to the capability of three-dimensional mapping local density variations, computed tomography (CT) offers excellent possibilities for non-destructive evaluation. The measured quantity is the averaged X-ray absorption over a volume element which size depends of the investigated sample and the experimental conditions.

The progress in spatial and density resolution, reached in the last years, together with the power of the three-dimensional imaging in times of a few hours or less makes the method suitable for a new starting point, resolving porosities, e.g. single pores and capillaries. The smallest volume element (Voxel) which can be distinguished from neighbor elements with a different density has an edge length of only a few μm. First experiments applying the computed tomography to determine the density distribution in high attenuation FGM samples like powder metallurgical parts are given by [1,2]. For powder metallurgy parts, the limited power of micro focal X-ray tubes and the beam hardening effect are a severe problem. For such cases correction algorithms have to be used for a quantitative image analysis [3,4]. For lighter materials like ceramic green parts, plastics and carbon, a simple filter can be used to narrow the energy spectrum of X-ray radiation without increase the measuring times to much.

## EXPERIMENTAL

Fabrication of porous ceramics

Ceramic powder was dispersed in aqueous solution at pH=6 and subsequently also with Na-alginate obtaining a slurry. The average particle size of ceramic powder is 180-350 nm depending of the ceramic material (measured by Microtrac Ultrafine Particle Analyzer; Leeds & Northrup). Then a solution of divalent metal ions ($Me^{2+}$) is deposited onto the surface of the slurry.

The slurry can be gelled by ion exchange of $Na^+$ in the alginate by divalent metal ions such as $Cd^{2+}$, $Pb^{2+}$, $Ca^{2+}$, $Zn^{2+}$ and $Sr^{2+}$. The gelation reaction is described in equation 1:

$$2 \text{ Na-alginate}_{(aq.)} + Me^{2+} \rightarrow Me\text{-alginate}_2 + 2 \text{ Na}^+ + x \text{ H}_2\text{O} \qquad (1)$$

Immediately a primary thin gel layer is formed. The primary gel layer has the function of a selective membrane e.g. to pass the metal ions ($Me^{2+}$) but not the ions (such as ions of the Na-alginate and ceramic) of the slurry. Due to diffusion control of $Me^{2+}$ transport through the membrane, the slurry gradually transforms to the gel resulting in the formation of capillaries in the direction of $Me^{2+}$ diffusion. The scheme of the gel structure used to prepare structured ceramics is shown in fig. 1. The surface of a gel together with the micrograph of sintered sample show the shrinkage during the production process (fig. 2). Three drying methods were used and gave satisfying results. The first method of obtaining samples without any cracks was supercritical drying (K850 Critical Point Drier). The second method was simply evaporating the solvent in air at 20 °C very slowly. Finally freeze-drying is a possibility to sublimate the crystallized solvent.

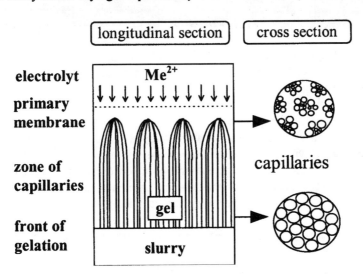

Fig. 1. Scheme of the gel structure used to prepare structured ceramics

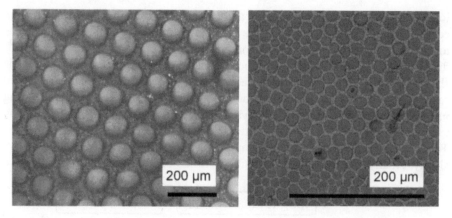

Fig. 2: The images of surface of a $Al_2O_3$ membrane gel sample (left half of image) and a micrograph of the sintered state (right half of image) show the shrinkage of the samples during the production process.

Setup of Computed Tomography

The high resolution tomograph developed at BAM is a cone beam type tomograph using an area detector that means the measured data at each angle increment are complete shadowgraphs or views. Since the emitted photons of the X-ray source form a cone beam, the scanned layers are not parallel. The angle of inclination rises with the distance from the central plane. Therefore special cone beam algorithms have to be developed for the image reconstruction. The most common used formalism is the Feldkamp [5] algorithm, which gives good results for opening angles of the cone beam lower than 15 degrees. Fig. 3 shows the apparatus with the transmission target X-ray tube (Feinfocus FXE-100.53, 100 kV, 1mA) at right. The focal spot is about 1-2 µm. As detector system (left side of fig. 3) a cooled CCD camera (Photometrix Series 200, CCD type TK1024 AF2 with 1024*1024 pixel, pixel size 24*24 $µm^2$) was used coupled to a 2:1 reducing fiber taper, which is coated with a GdOS scintillator screen of thickness of about 25 µm. The read-out rate of the CCD camera is 200 kHz with 16 Bit. The magnification, that means the ratio of the distance detector source to the distance object source can be adjust up to a factor of 100 (only for small objects with an diameter of less than 1 mm). Objects up to a diameter of 15 mm can be measured.

The result of a CT measurement is usually given in form of a 3D image matrix. Each point represents a single volume element. CT delivers a measure, the linear attenuation coefficient µ, for the absorbed X-ray radiation averaged over one voxel. The time needed for a complete measurement ranges from 30 min at minimum to several hours depending on the binning factor of the CCD camera, the selected Signal to Noise ratio and the attenuation through the sample. The reconstruction time is of the same order.

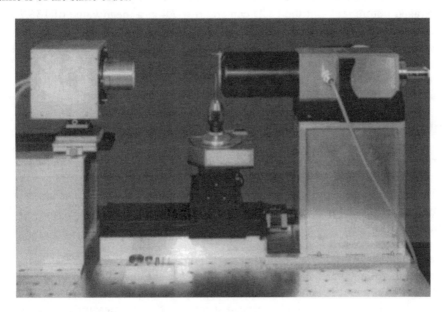

Fig.: 3: High-resolution cone beam tomograph developed at BAM. The right part shows the 100 kV transmission target X-ray source and the left part the detector system consisting of a scintillator coated fiber taper coupled to a CCD camera. A drill chuck is used as a sample holder.

Two important features characterizing the system performance of a tomographic apparatus are the spatial resolution and the contrast or density resolution [6]. The spatial resolution is usually defined as the 10% value of modulation transfer function (MTF). The experimental procedure determining the MTF is to measure a sharp edge of a homogenous material. From the edge response function (ERF) the MTF can be derived directly by a Fourier transform, without calculating the line spread function (LSF), which leads to an enhanced noise, as an intermediate step [7]. With a pixel size of $1.1*1.1*1.1$ $\mu m^3$ the ten percent value is about 320 Lp/mm corresponding to a spatial resolution of about 1.5 $\mu m$.

The standard deviation $\sigma$ of the contrast resolution $\Delta\mu/\mu$ which is defined as the FWHM value of the histogram of a homogenous sample was determined according the ISO norm from an area of $10*10$ pixel of a tomogram of a $SiO_2$

sample. For binning 1 of the CCD camera (1024 x 1024 image matrix) the standard deviation is $\sigma = 11.4$ grey levels at a mean value of 110.0 corresponding to a contrast resolution of $\Delta\mu/\mu = 10.3$ % and for binning 2, that means 4 pixel are averaged before read-out the CCD camera, $\sigma = 7.4$ at a mean value of 155.7 and $\Delta\mu/\mu = 4.8$ %. That means the photon statistic is the limiting factor for contrast resolution. These measurements are performed with an X-ray energy of 50 kV, current of 160 μA and prefiltering the X-ray spectra with 3 mm Al. The integration time was 50 sec per image without read-out time. At binning 1 a gain factor of 2 was selected to reduce the digitizing noise.

**RESULTS**

Bulk density of the ceramics was investigated by using the Archimedean principle. The tensile splitting strength (Instron Model 4301) was measured by diametrical test, also called the "Brazilian test".

The porosity and pore size distributions of the resulting porous ceramics was investigated by a mercury porosimeter (Shimadzu, Micromeritics Autoscan 33) and SEM (Digital Scanning Microscope 960, Zeiss). Furthermore shrinkage of samples was measured.

The shrinkage after the ion exchange, air-drying and sintering process shows the very high volume change of the samples related to the cut gel of up to 98 vol. %. The samples have a diameter of approximately 10 mm and a height of 3-6 mm. The ceramics have a bimodal pore distribution that is measured by mercury porosimeter. Fig. 4 shows the pore size distributions of the porous structured ceramics depending on the drying method. The first peak (pores radius between 100 and 1000 nm) represents the pores between the capillaries, the second (capillary radius between 2000 and 10000 nm) is caused by the capillaries themselves. The pores between the capillaries almost disappear with increasing sintering temperature. The uniform conical capillaries have a pore diameter of approximately 10-30μm in dependence of ceramic material and metal ions.

Fig. 5 and 6 show typical structures of a $TiO_2$ and hydroxylapatite sample. With increasing sintering temperature the bulk density and the tensile splitting strength increase whereas the porosity decreases.

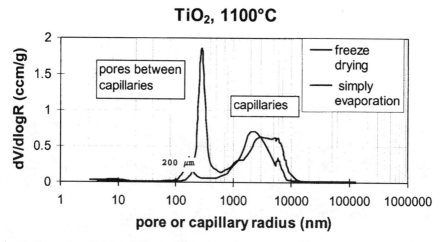

Fig. 4. Pore size distribution of a $TiO_2$ sample, measured by mercury porosimetry, as function of drying method.

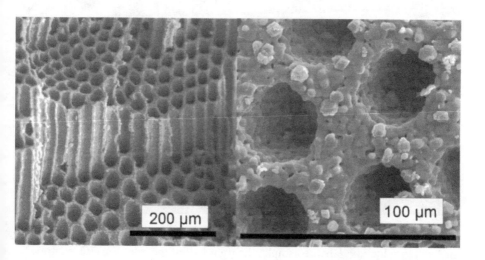

Fig. 5. Surface of fracture of a $TiO_2$ Sample (SEM)

Fig. 6. Surface of a hydroxylapatite sample (SEM)

Results of the CT measurements are shown in fig. 7 for a hydroxylapatite sample (diameter of about 7 mm) and a TiO$_2$ sample (diameter of about 6.5mm). The scan conditions are an image matrix of 907*907*297 and voxel size of 8*8*8 $\mu m^3$ and an image matrix of 937*937*486 and voxel size of 7*7*7 $\mu m^3$ respectively. The used maximum X-ray energy was 80 kV, 100 $\mu$A and a filter of 0.2 mm Cu. 720 projections were measured.

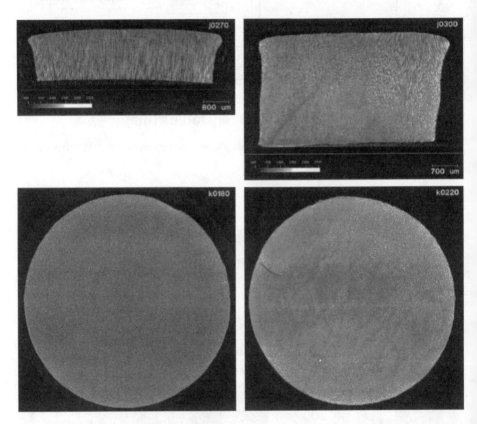

Fig. 7: Vertical and horizontal cross-section from the 3D image data set. The left side shows a hydroxylapatite sample, the right half of the image a TiO$_2$ sample.

To analyze more precisely the pore size distribution, the diameter of the TiO$_2$ sample was reduced to 3.3. mm and a corresponding voxel size of 3.6*3.6*3.6 $\mu m^3$. The selected X-ray conditions were a maximum X-ray energy of 50 kV, 180 $\mu$A and a filter of 2 mm Al. Fig 8 shows the pore size distribution for eight slices of this sample after an image analysis. The distance of the analyzed slices was 360 µm.

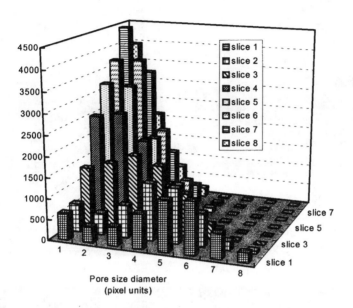

Fig. 8: Pore size distribution in a $TiO_2$ sample determined by CT. The distance between the analyzed slices is 360 µm.

From top to the bottom of the $TiO_2$ sample the pore size diameter decreases from about 6 pixel units (22 µm) to less than 1 pixel unit (3.6 µm).

The second feature of 3D image analysis is the search for throughput of capillaries. As example fig 9a and 9b shows the starting line in an image sector where a search algorithm developed at BAM was applied. The capillaries are visualized in fig. 9b. Due to the effect that some capillaries touch another the search algorithm finds out capillaries going back to surface shown as red spots beside the line.

A further reducing of the sample size increase the spatial resolution down to $2*2*2$ µm³. For this case even the pores in the walls between the capillaries could be resolved.

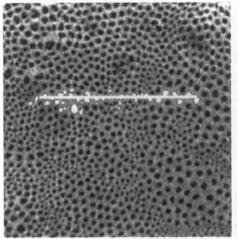

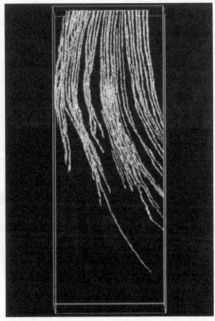

Fig. 9a and b: Searching for throughput of capillaries. The white line shows the starting area for search algorithm. The result is shown in fig. 9b (right half of image)

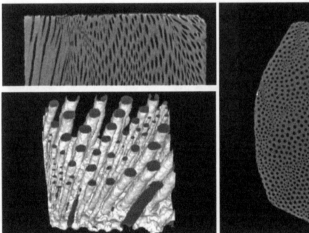

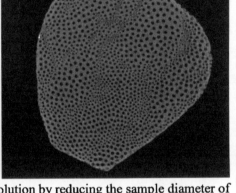

Fig. 10 Further increasing of spatial resolution by reducing the sample diameter of the TiO$_2$ sample. The pixel size is $2*2*2$ $\mu m^3$. The picture shows a vertical and horizontal slice of the sample together with a 3D visualization of some pore channels.

## CONCLUSION

With mercury intrusion porosimetry the pore size distribution averaged over the whole sample could be determined. Such measurements contains no information about local differences, which are usual due to the time-dependent behavior of pore size growth and clarified by REM images.

For the quantitative characterization of the gradient of porosity in direction of growing capillaries a set of micrographs was required to determine the following dependencies:

- pore size distribution in the x-y plane
- change of pore size in z-direction
- morphology of pores
- pore density as function of space

The basis was methodological work for metallographic preparation of plan parallel slices and their image analysis. Some difficulties are given by the porosity in the walls and the small contrast between ceramic phase and the pores infiltrated with organic material. Therefore the number of areas was limited and a complete automatic analysis could not be performed.

The advantage of CT is the non-destructive three-dimensional characterization of samples together with the capability for an automatic image analysis. Pore size distributions as well like geometrical features as the throughput of capillaries could be determined.

Further improvements in the detector system technologies - CCD cameras with up to 4096 * 4096 pixels are available - together with a reduction of costs should allow that larger samples could be characterized at the same spatial resolution as the above described samples. Alternative methods limited to a sector of a sample are the Region-of -Interest technique where only a part of a sample would be measured with high resolution whereas the whole sample was measured with coarse spatial resolution. Such a scanning and reconstruction technique was successfully developed for the two-dimensional case [8] and has to be extended for cone beam tomography.

## ACKNOWLEDGEMENT

The work was supported by the Deutsche Forschungsgemeinschaft (DFG) as part of the German FGM project and by DFG project „Structural Ceramics" in cooperation of the Freiberg University of Mining and Technology, Institut of Ceramics Materials (Prof. Tomandl) and the University of Regensburg, Laboratory of Surface and Chemistry, Institute of Analytical Chemistry, Chemo- und Biosensors (Prof.Heckmann)

We are grateful to our co-workers at BAM E. Jasiuniene and B. Illerhaus for developing algorithms for image analysis.

## REFERENCES

[1] H. Riesemeier, J. Goebbels, B. Illerhaus, Y. Onel, P. Reimers. "3-D Mikrocomputertomograph für die Werkstoffentwicklung und Bauteilprüfung", *DGZfP BB* 37 (1993) S. 280-287

[2] J. Goebbels, B. Illerhaus, G. Weidemann, K. Pischang
Characterization of Volume Properties of Powder Metallurgical Parts by Computerized Tomography
*Materials Science Forum* Vols. 308-311 (1999)pp. 867-872
1999 Trans Tech Publications, Switzerland

[3] E. Segal, W.A. Ellingson,Y. Segal, and I. Zmora, "Development of Beam Hardening Correction Method for Computerized Tomographic Imaging of Structural Ceramics: ", in: *Review of Progress in Quantitative NDE*, Vol. 6A, D.O. Thompson and D.E. Chimenti, es., Plenum Press, New York (1987), pp.411-417

[4] B. Illerhaus, J. Goebbels, O. Haase, H. Riesemeier, A.V. Kulish, M. Sené, M. Bailey
"Beam hardening correction in CT with simulated and real object data"
Proceedings of 2$^{nd}$ Int. Conf. on Computer Methods and Inverse Problems in Nondestructive Testing and Diagnostics, Minsk 20-23 October 1998, *DGZFP BB* 64 (1998), pp. 67-75

[5] L.A. Feldkamp, L.C. Davis, and J.W. Kress, "Practical Cone-beam algorithm", *J. Opt. Soc. Amer.* Vol. 1 (1984)pp. 612-619

[6] ISO 15708 Parts 1 and 2
*NDT-Radiation Methods-Computed Tomography-Part 1: Principles*
*NDT-Radiation Methods-Computed Tomography-Part 2:Examination Practices*

[7] N.J. Schneider and S.C. Bushong
Single-step calculation of the MTF from the ERF
*Med.Phys.* 5 (1978)pp. 31-33

[8] P. Reimers, A. Kettschau and J. Goebbels
"Region-of-interest (ROI) mode in industrial X-ray computed tomography"
*NDT International*, 23 (1990), pp. 255-261

# NOVEL PROCESSING FOR COMBINED COATINGS WITH DRY LUBRICATION ABILITY

R. Gadow, D. Scherer*
University of Stuttgart
Institute for Manufacturing Technologies of Ceramic Components and Composites (IMTCCC/IFKB)
Allmandring 7b,
D-70569 Stuttgart, GERMANY

ABSTRACT

Due to economical and ecological considerations there is an increasing demand for the use of light weight materials like magnesium and aluminum alloys. However, a major drawback of most light metals are their poor surface properties regarding corrosion resistance and tribological behavior. It is therefore essential to apply functional coatings for an additional surface protection. This presentation focuses on the deposition of tribologically advantageous combined coatings. In this approach thermally sprayed metallurgical or ceramic primary coatings with a high hardness and wear resistance are combined with a successive top coating with dry lubrication ability. In this study solid lubricant top coatings containing DLC, $MoS_2$, PTFE and other dry lubricants are deposited with various methods like chemical and physical vapor deposition (CVD, PVD) or lacquer spraying. The combined coatings show dry lubricant properties and lead to friction coefficients in the range of 0.1 under dry sliding conditions. In addition, they also provide excellent wear resistance and compressive strength.

INTRODUCTION

For many industrial applications there is a continuous demand for lightweight engineering materials. Especially in the automotive industry lightweight alternatives replacing steel are needed to realise weight savings in future products without sacrificing the room and performance that motorists are looking for. Industry engineers are looking at a variety of alternatives, including lightweight metals, plastics and various composite materials. All of these materials however pose a variety of challenges, including cost and manufacturability. The use of aluminum alloys is well positioned, but it is mainly applied in components like suspension pieces, engine blocks or small body panels. Magnesium alloys are expected to see exponential growth over the next years, although its use is quite limited today. The Ford Motor Co. projects that in 20 years its average vehicle will use 113 kg of the metal, which is up from barely 2.3 kg today [1].

A more widespread use of light metal alloys in tribological applications like guide bars, bearing plates, seat supports or bushings demands powerful functional surface coatings to provide wear protection as well as compressive strength. Direct contacts of uncoated light metal substrates with sliding or oscillating counterparts result in severe wear, seizing and high friction coefficients, even under lubricated conditions. There are different surface treatment processes for aluminum and magnesium alloys commercially available to increase the corrosion resistance and the wear resistance, like anodic oxidation or electroless nickel plating with codeposited PTFE and silicon carbide particles. But the use of these technologies does not match the requirements of many tribological applications, e.g. sliding movement under very high surface loadings. Another issue are the relatively high costs and the process technology required for the existing surface treatment processes.

An important driving factor for new developments in the field of tribology are environmental considerations concerning the effects of lubricants and grease on the ecological systems. Most lubricants contain environmentally harmful chemical additives. A significant amount of these lubricants is released to the environment, either by purpose or by accident. Therefore there is a steady demand for materials and surface coatings with solid lubricant ability and dry friction capability. By the use of these systems either no lubricants at all are needed or the amount of lubricants used can be reduced drastically.

In this paper the concept and the processing of combined coatings on light metal alloys is introduced. The combined coatings are designed to provide a high wear resistance and high compressive strength for the light metal substrate as well as to exhibit dry lubrication properties. The combined coatings consist of a wear resistant metallurgical, ceramic or cermet primary layer which is applied on the light metal substrate by thermal spraying. Successively, a secondary coating layer with dry lubrication ability is applied to provide a low coefficient of friction. This secondary layer is either a polymer based lubricating varnish which consists of a polymer matrix containing finely dispersed solid lubricants (e.g. PTFE, $MoS_2$, graphite, $BN_{hex}$) or it is applied by thin film technology, i.e. by physical vapor deposition (PVD) or by chemical vapor deposition (CVD). In this study a lubricating varnish containing PTFE particles as solid lubricants, diamondlike carbon (DLC) coatings deposited by plasma enhanced chemical vapor deposition (PECVD) and nanostructured molybdenum disulphide ($MoS_2$) coatings deposited by cathode sputtering (DC magnetron sputtering) were used as solid lubricant coatings on the thermally sprayed primary layers.

PROCESSING OF COMBINED COATINGS ON LIGHT METAL SUBSTRATES

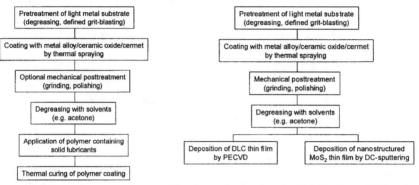

Fig.1 Fabrication routes for combined coatings with dry lubrication ability

In Fig. 1 the different fabrication routes for the combined coating systems are shown. The wear resistant metallurgical, ceramic or cermet primary layer is applied by thermal spraying for all different coating systems. The thermal spray technology is well established and allows the application of a broad variety of materials in form of spray powder on different substrate materials. Most thermal spraying techniques use combustion gases and plasma as sources of the thermal and kinetic energy heating the powder particles and propelling them to the substrate to coat. Prior to the application of the thermally sprayed layer, the light metal substrates are degreased and grit blasted to increase the surface roughness, since the adhesion of the thermally sprayed layer is mainly caused by mechanical

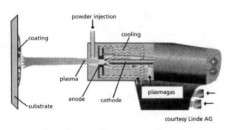

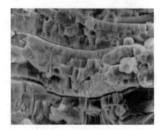

Fig. 2 Plasma torch for Atmospheric plasma spraying (APS)

Fig. 3 SEM micrograph of lamellar coating build-up of APS sprayed $TiO_2$

clamping. The molten particles which strike the substrate surface deform to lamellae and solidify. Due to the shrinkage of the particles during cooling down they are mechanically "anchored" to the irregularities of the substrate, i.e. to the surface asperities. In this study atmospheric plasma spraying (APS) was used to apply the wear resistant primary coatings. Details of the thermal spray technology can be found in the literature [2]. In Fig. 2 a schematic drawing of a torch used for atmospheric plasma spraying is shown and Fig. 3 shows a SEM micrograph of the typical lamellar coating build-up of an oxide ceramic deposited by APS. The typical surface roughness of a thermally sprayed coating is in the range of approx. 20 to 60 μm.

*Combined thermal spray/polymer coatings:* For most combined thermal spray/polymer coatings the typical surface roughness and topography of thermally sprayed layers is beneficial since the "valleys" act as a reservoir for the lubricating varnish which is constantly released during operation. This characteristic has been shown in other studies before [3]. In case the surface roughness is too high after thermal spraying, an optional posttreatment step might be added to remove the most protruding asperities by mechanical grinding and polishing and to create a more homogeneous load bearing area.

Table 1 Typical composition of a lubricating varnish

| *Compound* | *Example* |
|---|---|
| Solid lubricant | Graphite, PTFE, $MoS_2$, $BN_{hex}$ |
| Polymer binder matrix | Epoxy resin, silicone resin, polyester resin, polyamide/polyimide resin |
| Solvent | Water; Alcohol, ester, ketone, glycol ester |
| Additives | Pigments, leveling agents, dispersing additives |

Different technologies can be used to apply the lubricating varnish, like dipping, brushing and rolling. For this study pneumatic air spraying was used to deposit the lubricating varnish. The selected coating thickness was about 30 to 50 μm, i.e. it is in the range of the surface roughness of the thermally sprayed layers. After application the coatings are thermally cured. In Table 1 the composition of a typical lubricating varnish is summarized.

Fig. 4 shows SEM micrographs of different solid lubricants. A lubricating varnish typically contains 20 to 40 vol. % solid lubricant particles to yield a low friction coefficient. In Figs. 5 and 6 SEM micrographs of metallographic cross sections of two different lubricating varnishes are shown which are applied on a thermally sprayed TiO$_2$ layer. Both varnishes were deposited on the oxide ceramic layer in the state "as sprayed", i.e. the TiO$_2$ coating was not grinded after spraying.

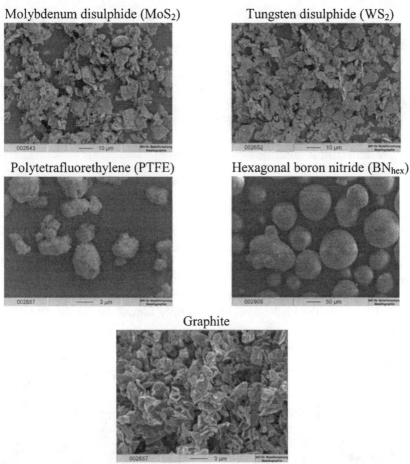

Fig. 4 SEM micrographs of solid lubricants used in lubricating varnishes

The wetting and adhesion of the varnish on the oxide ceramic coating is very good. All "valleys" are filled with polymer and the surface asperities are also well covered. Upon tribological load, i.e. during the "running-in" period, excess polymer is removed and the surface is homogenized. The TiO$_2$ coating forms the main load bearing area and provides wear resistance and compressive strength, whereas the lubricating varnish is constantly released out of the "valleys" and deposited between the sliding counterparts resulting in a low coefficient of friction. The lubricating varnish shown in Fig. 5 contains only PTFE particles as solid lubricant, whereas the one shown in Fig. 6 contains PTFE as well as MoS₂ particles.

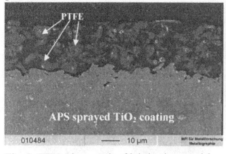
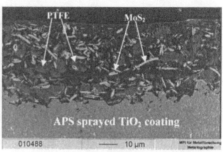

Fig. 5 SEM micrograph of lubricating varnish containing PTFE particles as solid lubricants on APS sprayed TiO₂ coating

Fig. 6 SEM micrograph of lubricating varnish containing PTFE and MoS₂ particles as solid lubricants on APS sprayed TiO₂ coating

*Combined thermal spray / thin film coatings:* For the processing of the combined thermal spray/thin film coatings, all thermally sprayed layers are first grinded and polished to a resulting surface roughness of $R_a$=0.05 µm and $R_z$=1.0 µm. The coating thickness of both the cathode sputtered MoS₂ as well as the DLC coatings applied by PECVD was 2 µm. Therefore an application of these thin films on the thermally sprayed primary layer in the state "as sprayed" is not successful, because the load bearing area is only formed by the most protruding asperities and no running-in is possible. This causes high specific pressures at the asperities which results in severe wear and a high coefficient of friction.

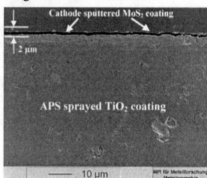

Fig. 7 Fracture morphology and surface topography of on MoS₂-based solid lubricant film with Cr intralayers (MoS₂ layers appear dark, Cr layers appear bright) [5]

Fig. 8 SEM micrograph of cathode-sputtered MoS₂ film on thermally sprayed TiO₂ coating

Innovative Processing and Synthesis of Ceramics

The first part of the samples was coated with a dense MoS$_2$-based solid lubricant film having a defined nanostructure to avoid a columnar microstructure. This coating was applied by a special cathode sputtering setup where tungsten or chromium intralayers were introduced by rotating the substrates between the MoS$_2$ and a tungsten or chromium target. Thus the MoS$_2$ film growth is interrupted repeatedly, resulting in a dense, non-columnar microstructure and a long durability of the coating. Details if this coating process can be found in the literature [4], [5]. For the experiments shown in this study, tungsten intralayers were introduced and a very fine multilayer stacking W/MoS$_2$ was selected. In Fig. 7 an example of such a nanostructured MoS$_2$–based thin film with Cr intralayers is shown and Fig. 8 illustrates a cathode sputtered MoS$_2$ coating on a polished, thermally sprayed TiO$_2$ oxide ceramic layer.

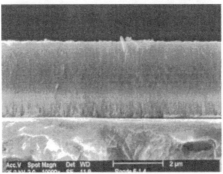

Fig. 9  Diamondlike carbon (Me-C:H) coating deposited by PECVD process [6]

The other part of the substrates was subject to two different DLC coating processes. DLC coatings are one type of carbon based hardcoatings exhibiting high hardnesses and low friction coefficients. Carbon based thin film coatings are deposited via several mixed PVD / CVD techniques in a vacuum chamber (typical pressures lie between $10^{-2}$ to $10^{-3}$ mbar). Hydrocarbons (acetylene) are introduced into the reaction chamber and are cleaved in a plasma. To deposit metal containing carbon based coatings (Me-C:H), PVD techniques are used to evaporate the metal (magnetron sputtering, random Arc). Pure carbon coatings (a-C:H) are deposited via plasma assisted CVD. In this study, pure carbon coatings (a-C:H) and titanium containing carbon based coatings (Ti-C:H) were evaluated. In Fig. 9 a SEM micrograph of a typical Me-C:H coating is shown. The coating exhibits a layer structure starting from a more crystalline metal rich phase near the substrate interface to a more amorphous, carbonrich phase towards the surface. For the experiments a coating thickness of 2 μm was selected.

## MECHANICAL AND TRIBOLOGICAL EVALUATION OF COMBINED COATINGS

Two light metal alloys were used as substrate materials, the magnesium alloy AZ91 and the aluminum alloy AlMg3. The magnesium alloy AZ91 with 9 % aluminum and 1 % zinc shows excellent castability, high strength and good corrosion properties and represents among the various types of magnesium cast alloys 90 % of the consumption, e.g. used for engine and transmission components. Typical material properties of the two light metal alloys are given in Table 2.

Table 2  Important physical and mechanical properties of magnesium alloy AZ91 and aluminum alloy AlMg3 [7], [8].

|  | Density $\rho$ [g/cm³] | Linear CTE $\alpha$ [$10^{-6}$/K] (273 K – 373 K) | Thermal cond. $\lambda$ [W/K*m] (293 K) | Ultimate tensile strength $\sigma$ [MPa] | Young's modulus E [GPa] |
|---|---|---|---|---|---|
| AZ91 | 1.81 | 26.0 | 51 | 240 | 45 |
| AlMg 3 | 2.66 | 23.8 | 134 | 240 | 70 |

Two oxide ceramic coatings, $Al_2O_3$ and $TiO_2$, were applied by APS on the two light metal substrates as primary coatings. In Table 3 selected properties of these coatings are summarized.

Table 3  Surface roughness $R_a$ and $R_z$ , porosity, universal hardness (HU), Vickers hardness (HV0.05) and bonding strength of APS sprayed coatings on AZ91

|  | Spray technology | $R_a$ [µm] | $R_z$ [µm] | Porosity [%] | HU [N/mm²] | HV0.05 [-] | Bonding strength [N/mm²] |
|---|---|---|---|---|---|---|---|
| $Al_2O_3$ | APS F4 | 3.9 | 27.9 | 6.5 | 7830 | 1252 | 16,0 |
| $TiO_2$ | APS F4 | 3.7 | 27.2 | 5.4 | 6136 | 962 | 16,9 |

The bonding strength of the thermally sprayed coatings on the substrate was determined by pull testing. The tribological evaluation of the combined coatings was carried out in a tribometer using an oscillating ball-on-disk setup. The oscillating mode was used, since due to the turning points very severe test conditions are created. A normal load of $F_N$=10 N was applied, the test velocity was 70 mm/s, the humidity was held constant at r.h.=40 %, the amplitude of oscillation was 5 mm and the tests were performed up to 100.000 oscillations. A 100Cr6 ball with a diameter of 5 mm was used as counterpart.

In order to compare the tribological performance of the combined coatings with the tribological performance of the uncoated light metal substrates, ball-on-disk tests were first carried out on AZ91 and AlMg3 substrates without any coatings. As a second test, the lubricating varnish and the different thin films were applied directly on the light metal substrates after the surfaces were polished and degreased, i.e. without a thermally sprayed primary coating. In Fig. 10 the dry sliding behavior of the uncoated light metal substrates as well as of the solid lubricant films which were directly applied on the substrates is shown. For the uncoated samples, the coefficient of friction increases rapidly for AlMg3 and AZ91 from μ=0.6 to μ=1.2 and from μ=0.4 to μ=0.8, accompanied by severe seizing. This demonstrates that both light metals are not suitable for tribological applications without a protective surface coating. But even if a solid lubricant film is applied on the substrates, the friction behavior does not improve significantly, with the exception of the DLC coating. Only during the initial oscillations the coefficient of friction remains low, but then it increases drastically to high values and severe seizing occurs, equivalent to the tribological behavior of the light metal substrates having no solid lubricant coating applied at all.

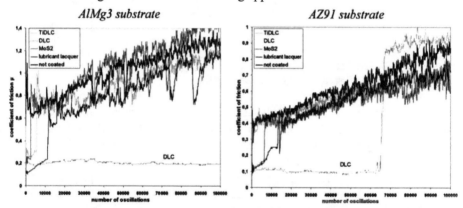

Fig. 10  Dry sliding behavior of uncoated AlMg3 and AZ91 substrates and of different solid lubricant films directly deposited on the light metal substrate without thermally sprayed primary coatings; oscillating sliding movement; counterpart: 100Cr6 ball

In Fig. 11 the dry sliding behavior of different combined coatings on a AlMg3 substrate are shown. An Al₂O₃ coating applied by APS was used as primary layer for all combined coatings. The polished Al₂O₃ coating itself shows a very high coefficient of friction against 100Cr6, but the addition of all solid lubricant films improved the dry friction behavior significantly. In particular, the coefficient of friction remains very low up to a high number of oscillations for the DLC and the TiDLC thin film coatings. Although the coefficient of friction increases to high values earlier if a lubricating varnish or a MoS₂ coating are used, the light metal substrate is still protected by the $Al_2O_3$ layer against seizing. The combined thermal spray/lubricating varnish coatings perform better if the $Al_2O_3$ layer is polished prior to the application of the lubricating varnish.

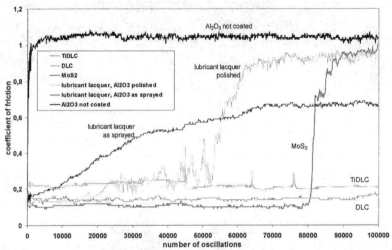

Fig. 11 Dry sliding behavior of different combined coatings on AlMg3 substrate against 100Cr6 ball with APS sprayed $Al_2O_3$ primary coating

Similar tribological characteristics under dry sliding conditions are seen for combined coatings on a magnesium substrate with a thermally sprayed $TiO_2$ primary layer, compare Fig. 12. In contrast to the $Al_2O_3$ based combined coatings, the combined $TiO_2$/lubricant varnish coating shows a better tribological performance if the lubricating varnish is applied in the state as sprayed. For the TiDLC coating several steep increases of the coefficient of friction can be observed followed by steep decreases. The increase is due to the local wear of the TiDLC coating at the sliding interface and as a result direct contacts between the 100Cr6 counterpart and the $TiO_2$ surface occur. Thus the 100Cr6 ball and the $TiO_2$ surface are locally worn and a wear scar on the 100Cr6 ball is formed causing an increase of the coefficient of friction. On the other hand, due to the formation of the wear scar on the 100Cr6 ball the load bearing area increases and contacts with "unconsumed" TiDLC at the sides of the wear track lead to a low coefficient of friction again.

For a tribological evaluation of the combined coating systems, not only the coefficient of friction has to be considered, but also the wear occurring on the surface of the combined coating as well as on the 100Cr6 counterpart. The volumetric wear values are summarized in Figs. 13a and 13b. It is obvious that the application of a solid lubricant films reduces the volumetric wear at the

Innovative Processing and Synthesis of Ceramics

100Cr6counterpart drastically. The combined coatings which show the lowest friction coefficient also show the lowest wear at the 100Cr6 counterpart. Care has to be taken to interpret and compare the surface wear rates. If the wear volume is measured by analysis of the wear track on the substrate surface, both the wear in the solid lubricant film and in the thermally sprayed primary coating are recorded. Thus the high volumetric wear values for the combined coatings with lubricating varnish are mainly due to the removal of excess varnish during the running-in period, so here the wear volume is mainly formed in the lubricating varnish.

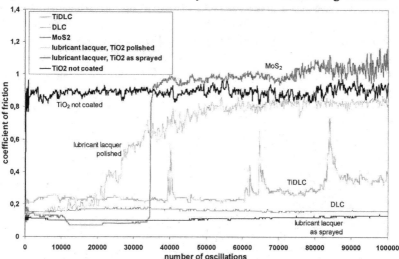

Fig. 12   Dry friction behavior of combined coatings on magnesium substrate against 100Cr6 ball with APS sprayed TiO₂ primary coating

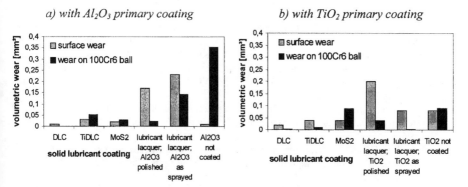

Fig. 13   Wear in combined coating systems at substrate surface and on oscillating 100Cr6 counterpart after 100 000 oscillations for different coating systems

## SUMMARY AND CONCLUSION

The concept of combined coatings can be effectively used to improve the friction and wear properties of light metal substrates under dry sliding conditions. Without protective coatings, light metal substrates show a poor tribological behavior which has been demonstrated in pin-on-disk testing using a 100Cr6 counterpart. As a result, high coefficients of friction and severe wear and seizing was observed. The deposition of various solid lubricant coatings directly on the light metal substrates did not significantly improve the tribological properties, because of the poor adhesion of some of the coatings and due to the low load bearing capacity of the substrates. The deposition of two oxide ceramic coatings, $TiO_2$ and $Al_2O_3$, as primary coatings by thermal spraying leads to improved wear resistance, compressive strength and load bearing capacity of the substrates. The additional application of solid lubricant films by PVD and CVD techniques and by polymer deposition results in low coefficients of friction and a good tribological behavior under dry sliding conditions. Even after the solid lubricant coatings were consumed and the coefficient of friction did increase to high values, no seizing occurred since the thermally sprayed primary coatings provide wear protection for the light metal substrate. The concept of combined coatings will open new fields for the use of light metal alloys in tribological applications and a significant decrease or even replacement of grease and fluid lubricants is possible in many engineering applications.

## REFERENCES

[1]   C. Winandy, „Editor's Note", in *Automotive Light Metals 1-2*, pp. 9, December 2000, First Global Media Group, New Jersey, USA

[2]   L. Pawlowski, *The science and engineering of thermal spray coatings*, John Wiley & Sons Ltd Chichester, 1995, ISBN 0 471 95253 2

[3]   Scherer, D.; Gadow, R.; Killinger, A., *Manufacturing and experimental evaluation of combined ceramic polymer coating systems for tribological applications*", UTSC, United Thermal Spray Conference 99, 17.-19. März 1999, Düsseldorf; Conference Proceedings; E. Lugscheider, P.A. Kammer (ed.), pp. 664-669; ISBN 3-87155-653-X

[4]   M.W. Nordbakke, F. Heutling, M. Meyer, O. Knotek, "New aspects regarding sputter-depositing dense coatings, in particular a solid lubricant of practical interest", *Mat.-wiss. u. Werkstofftechn.* 31, pp. 205-214 (2000), WILEY-VCH, Verlag GmbH, D-69451 Weinheim, 1998, ISSN 0933-5137 (in German)

[5]   M.W. Nordbakke,"Nanostrukturierte Festschmierstoffschichten auf $MoS_2$-Basis", Fortschrittsberichte VDI-Reihe 5 Nr. 586, Düsseldorf, VDI Verlag (2000); ISBN 3-18-358605-3

[6]   J. Brand, R. Gadow, A. Killinger, "Diamondlike and hydrogen-carbon hardcoatings for tools in the manufacturing of high precision glass components, Proceedings of the Glass Processing Days, 13.-16. Juni 1999, Tampere, Finnland

[7]   T. Merkel, *Taschenbuch der Werkstoffe*, Fachbuchverlag Leipzig, Germany, 1994

[8]   "Die Cast Magnesium Alloys", Hydro Magnesium, Data Sheet, 1997

# CERAMIC COATINGS ON FIBER WOVEN FABRICS FOR LIGHT-WEIGHT BALLISTIC PROTECTION

Konstantin von Niessen and Rainer Gadow
University of Stuttgart
Institute for Manufacturing Technologies of Ceramic Components and Composites (IMTCCC/IFKB)
Allmandring 7b
D-70569 Stuttgart, GERMANY

## ABSTRACT

Based on thermal spray technologies a coating process for refractory oxide ceramic layers even on temperature sensitive fiber substrates has been developed, so that the coated fabrics retain their flexibility. High speed and high rate ceramic coating is performed with simultaneous cooling so that refractory oxide ceramic coatings can be applied on aramide and mullite fibers with potential for industrial application. The penetration by bullets, knives and blades through such ceramic coated multilayer fabrics is effectively prevented.

## INTRODUCTION

Ballistic protection is required for personal use, vehicles and permanent structures which are subject to ballistic threats.[1,2] Military and civilian ballistic protection is divided into flexible light weight protection and massive, stiff armor. Light weight ballistic protection is made of flexible aramide fibers and is primarily used as body armor[3,4] (see Fig. 1).[5] Stiff armor consists of multilayer steel as well as dense bulk ceramic plates and stiff fiber reinforced materials (see Fig. 2).[6] The main disadvantages of solid and stiff armor are the high weight and the inflexibility. Because of the stiffness, hinges and hatches cannot be protected. For personnel protection as well as protection of aircrafts and cars, only light and flexible materials can be used.[7] Light and flexible fabrics made of aramide or other high tenacity fibers meet some of these demands but their protection is not sufficient. Sharp blades as well as high speed bullets can pierce these fabrics even if several layers are used. The state of the art solution to protect body armor against knives and bayonets is the extra use of layers made of titanium foil (see Fig. 3) which significantly increases the specific weight.[8] This paper focuses on a new approach by coating fabrics made of high tenacity fibers such as aramide and mullite fibers with a highly refractory oxide ceramic by thermal spray technologies. By the combination of high tenacity fiber woven fabrics and high performance ceramic coatings the penetration by bullets, knives and blades can be effectively prevented. The ceramic coating increases the fiber to fiber friction which prevents wave distortion and delamination. The penetrating objects cannot change the fabric structure and push the fibers aside. The hard oxide ceramic coating blunts sharp metal blades by abrasion so they cannot trench the fabric, and the high friction between the ceramic coating and the metal blade stops further penetration.

**Fig.1** Personnel protection made of fiber fabrics[8]

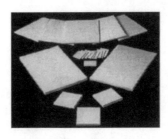

**Fig.2** Heavy armor made of bulk ceramic[5]

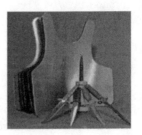

**Fig.3** Extra protection made of titanium foil[8]

MATERIAL SCREENING

The material screening focusses on the use of high tenacity fiber fabrics and highly refractory oxide ceramics. Two different commercially available fiber fabrics have been selected, the standard aramide fabric used for ballistic protection Twaron[©] (Twaron Products, Wuppertal, Germany) and the mullite fiber fabric Nextel[©] 720 (3M, Minneapolis, MN, USA) consisting of 85% $Al_2O_3$ and 15% $SiO_2$. The material properties of these fibers are summarized in table I.

Table I. Properties of fiber fabrics[9]

| Fiber fabric | Density $\rho$ [g/cm$^3$] | Tenacity $\sigma$ [MPa] | Initial modulus E [GPa] | Decomp. temp. $T_D$ [°C] | Specific heat $C_P$ [J/kgK] | Max. appl. Tem. $T_M$ [°C] |
|---|---|---|---|---|---|---|
| Twaron[©] | 1.45 | 2,800 | 85 | 500 | 1420 | 200 |
| Nextel[©] 720 | 3.40 | 2,100 | 260 | 2,000 | 800 | 1,204 |

Due to their high hardness and wear resistance the oxide ceramics $Al_2O_3$ and $TiO_2$ have been chosen as coating materials for thermal spraying. To improve the bonding strength of the ceramic coatings on the fabric, AlSi is used as additional bond coat. The bulk material properties of the ceramic materials are shown in table II.

Table II. Bulk material properties of $Al_2O_3$ and $TiO_2$[10]

| Oxide ceramic | Density ρ [g/cm³] | Vickers hardness HV [-] | Youngs modulus E [GPa] | Melting temp. $T_M$ [°C] | Specific heat $C_P$ [J/kgK] |
|---|---|---|---|---|---|
| $Al_2O_3$ | 3.98 | 2,200 | 400 | 2047 | 1,047 |
| $TiO_2$ | 4.25 | 1,150 | 205 | 1,860 | 730 |

In order to apply these oxide ceramics by thermal spraying, they have to be available as spray powders. After a sintering process the used powders are mechanically broken and milled to a grain size of 10 – 22 μm.

DEPOSITION OF OXIDE CERAMIC COATINGS ON LIGHTWEIGHT FIBER FABRICS BY THERMAL SPRAYING

The thermal spray process allows the application of a broad variety of metallurgical, cermetic and ceramic coatings on a variety of substrates. A key feature of the thermal spray technique compared to many other methods is the substrate's low thermal load during the coating process. By using simultaneous air or liquid $CO_2$ cooling techniques, the substrate's temperature can be kept relatively low e.g. between 50° and 150°C. The Atmospheric Plasma Spray (APS) process uses as an energy source an electric arc discharge between a water cooled copper anode and a tungsten cathode. This electric arc discharge dissociates and ionises the working gas and builds up a plasma that expands into the atmosphere forming a plasma gas jet (see Fig.4).[11]

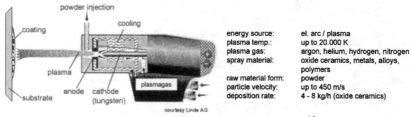

| | |
|---|---|
| energy source: | el. arc / plasma |
| plasma temp.: | up to 20.000 K |
| plasma gas: | argon, helium, hydrogen, nitrogen |
| spray material: | oxide ceramics, metals, alloys, polymers |
| raw material form: | powder |
| particle velocity: | up to 450 m/s |
| deposition rate: | 4 - 8 kg/h (oxide ceramics) |

**Fig. 4** The Atmospheric Plasma Spray (APS) process[12]

The spray powder, suspended in a carrier gas, is injected into the heat source of the torch. After being totally or partially molten and being accelerated, the powder particles impact on the substrate's surface. There they are quenched and solidified within $10^{-5}$ to $10^{-7}$ seconds. The coating buildup is the result of the molten powder particles impacting one upon the other (see Figs. 5 and 6).

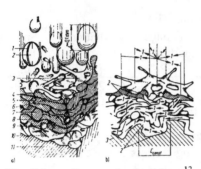

**Fig. 5** Lamellar coating build up [13]

**Fig. 6** SEM of a $TiO_2$- coating structure

During atmospheric plasma spraying process temperatures up to 20,000 °C are obtained. Therefore this process is mainly used for deposition of refractory materials like oxide ceramics.

For the application of thermally sprayed coatings on fiber woven fabrics, two different manufacturing methods are considered. Depending on the spray distance and the used spray equipment, the plasma torch covers a strip with a width of approximatly 10 mm. In order to coat a larger surface a defined movement between torch and fabric has to be found. This movement can be performed either by the torch on a meandering feed or by rotating the fabric on a round table. Using a rotating table is particularly suited to coat several small samples within one coating process (see Fig. 7).

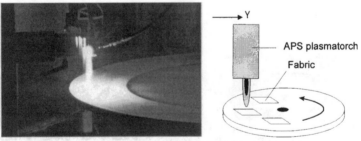

**Fig. 7** Rotating table arrangement for APS spraying

In order to coat individual fabric samples, the meandering feed arrangement is more suitable because the coating track can be adjusted to the size of the sample. The torch movement is performed by a xy- feed drive and a metal frame is used to insert and tenter the samples. A steel wire cloth within the frame supports the flexible fabric structure during the coating process and allows cooling because of its open and permeable design. To ensure identical spray parameters, both evaluated fabrics were inserted into the metal frame and coated together within one coating process. The meandering movement and the metal frame are shown in Fig. 8.

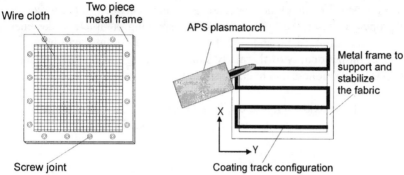

Fig. 8 Mounting support for the fabrics and coating track configuration

In order to limit the thermal load on the fabrics a simultaneous cooling with compressed air is used. Air nozzles are attached on both sides of the spraying torch. In addition, the process is supervised by an infrared camera (Varioscan InfraTec ID, Dresden, Germany) and in that way the temperature of the coated samples can be controlled in real time. Fig. 9 shows a typical IR- picture during the coating process.

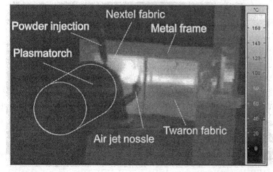

Fig. 9 IR- picture of the temp. distribution during the coating process

MECHANICAL CHARACTERIZATION

With regard to the use of the coatings for ballistic protection, the main focus of the characterization is on the determination of hardness and wear resistance as well as on the evaluation of the coating's bonding strength on the first fiber layers. During the coating buildup of thermally sprayed layers, porosity and microcracks cannot be avoided. For the coating of flexible fabrics the formation of porosity and microcracks in the coating is desired because it leads to a higher flexibility of the fabric. But if the porosity is too high, the hardness and other mechanical properties of thermally sprayed coatings decrease. So a balance between porosity and mechanical properties has to be found. Tuning the spraying parameters like energy supply, kinematic arrangement and simultaneous process cooling, will have a significant effect on the quality of thermally sprayed coatings with regard to microstructure, porosity, hardness and bonding strength.

The thickness of the oxide ceramic coatings on the fabric is in the range of 50 – 100 μm. Fig 10 shows a schematic drawing of the intended structure of the coated fabric.

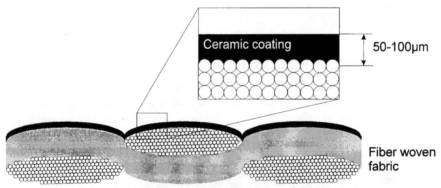

**Fig. 10** Intended structure of the oxide ceramic coated fabric

In Fig. 11 a cross section of a Twaron© fabric coated with an $Al_2O_3$ oxide ceramic layer is shown. The lamellar structure and the good wetting behavior of the ceramic coating on the first layers of the fabric are visible. The macro-structure and micro-structure of the coated fabric's surface is typical for thermally sprayed coatings (see Fig. 12). The structure of the fabric is still visible in the macrostructure. Even though the $TiO_2$- and $Al_2O_3$- coatings have melting points above 1800° and 2000°C respectively, there is no significant polymer fiber damage.

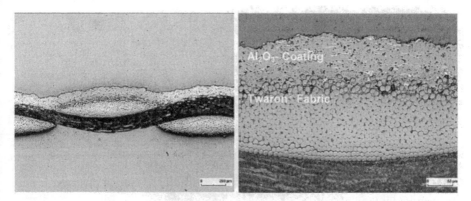

**Fig. 11** Cross section of a thermally sprayed Al₂O₃ coating on a Twaron© fabric

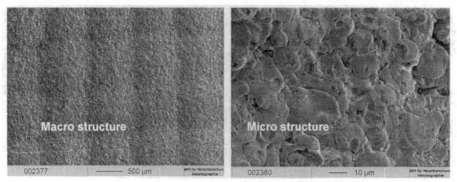

**Fig. 12** SEM micrographs of a thermally sprayed Al₂O₃ coating on a Twaron© fabric

In order to evaluate the coating quality metallographic examinations have been performed. The coating porosity determined by an image analysis is expressed by the relative pore volume content $V_P$ [%]. An automized universal hardness indenter equipment (Fischerscope TM HCU) with a load of 500 mN is used to determine the coating hardness HV0,05. In order to measure the hardness of an individual fiber, the load was reduced to 10 mN (HV 0,001). Table III and table IV show the measured porosity and hardness characteristics of the thermally sprayed coatings and of the fibers, respectively.

Table III. Measured coating porosity and hardness (HV 0,05)

| Coating | $V_P$ [%] | HV 0,05 |
|---------|-----------|---------|
| $Al_2O_3$ | 5.8 | 1,240 +/- 300 |
| $TiO_2$ | 3.2 | 1,100 +/- 110 |
| $Al_2O_3/TiO_2$ | 4.1 | 1,025 +/- 180 |
| AlSi | 1.44 | 138 +/- 10 |

Table IV Microhardness of individual fibers (HV 0,001)

| Fiber | HV 0,001 |
|-------|----------|
| Twaron[©] | 51.52 +/- 7 |
| Nextel[©] 720 | 1,610 +/- 405 |

One of the intentions of ceramic coatings on fiber fabrics is to blunt metal blades or other penetrating objects by abrasion. In order to judge the wear behavior of a metal counterpart on the oxide ceramic coatings, dry running oscillating pin on disc tests are performed. As a counterpart 100Cr6 balls with a hardness of 1.165 HV0.05 and a diameter of 5 mm are used. The selected number of oscillating strokes is 10.000, the sliding velocity is 70 mm/s, the length of strokes is 5 mm with an imposed normal load of 10 N. It can be assumed that a higher volumetric loss on the 100Cr6 ball after the end of the tribological tests is an indication for a better ability of the coating to blunt a metal blade. In order to estimate the abilities of the fabric itself to blunt a metal blade, non coated fabrics have also been tested. Fig. 13 shows the volume loss of the 100Cr6 balls for different coatings on both fabrics.

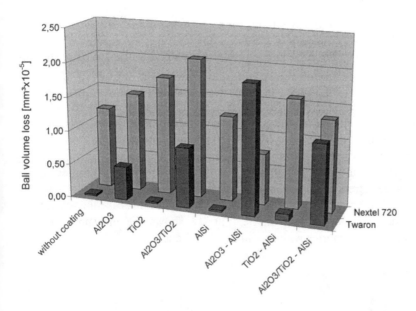

**Fig. 13** Volume loss measured on a 100Cr6 ball after 10.000 strokes in contact with coated fiber fabrics

The fiber fabric itself has a certain effect on the abrasion properties. Due to the high hardness, the Nextel fibers cause quite significant wear on the 100Cr6 ball even without any coating. In comparison to the Nextel fiber, Twaron is a rather smooth fabric which, without coating, does not cause any severe wear. The $Al_2O_3$ and $Al_2O_3/TiO_2$ coatings on Twaron cause a significant volume loss, which is further increased by using an additional AlSi bond coat. But since the microhardness of the AlSi bond coat is low and thus it does not have a noticeable impact on the abrasion behavior, only the increased higher bonding strength of a multilayer coating explains this result. Because of the protective function of the bond coat, the hard oxide ceramic coatings cannot easily be sheared. The comparatively hard $TiO_2$ coating has no significant influence on the volume loss, even with the AlSi bond coat. This might be due to the solid lubricant properties of $TiO_2$ which have been described in other investigations.[14-16]

The results of the tests with Nextel fabrics are different. All oxide ceramic coatings increase the wear volume on the ball. This increase is not only caused by the hardness of the coatings, but also is due to the fixation of the fibers by the coating. Nextel fibers usually can't bear lateral forces because there is little friction between them. By the deposition of a coating, the hard fibers are fixed. The additional AlSi bond coat seems to have no influence on this mechanism.

For industrial application the coating must also show a sufficient bond strength to the fabric. The investigation of the coating`s adhesion on the fabric is performed on a Zwick Z100 universal mechanical testing machine by pull testing. The coated fabric samples are glued to a metal plate and a steel tension rod is glued to the coating surface by using an adhesive. After mounting the samples into the testing machine the tension load is continuously increased. As soon as a delamination of the coating occurs, the tension load is measured and the bonding strength is determined. As the bonding strength of the coatings is limited by the maximum shear strength of the first fiber layers which are in contact with the coating, the fabrics are also tested without any coating. In this case the tension rod is glued directly on top of the fabric. Fig. 14 shows the measured bonding strengths for the used fabrics with or without AlSi bond coat.

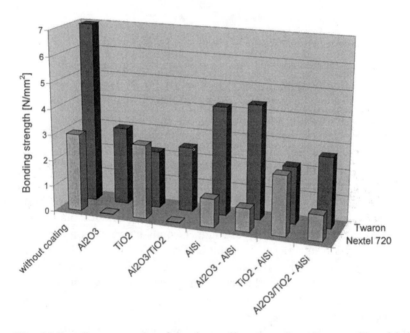

**Fig. 14** Bonding strengths of the thermally sprayed coatings on fiber fabrics

The results of the experiments with non–coated fabrics show the maximum possible bonding strength a coating could reach on the fabrics. Because of its low shear strength, Nextel already reaches its limit at a bonding strength of 3 N/mm².

The $TiO_2$ coatings reach this values with and without a AlSi bond coat. The bonding strength of the $Al_2O_3$ and $Al_2O_3/TiO_2$ coatings is rather low, however it can be increased by using the additional AlSi bond coat. For the $TiO_2$ coated Nextel fabrics delamination occurs within the fabric itself, whereas the other coatings with lower bonding strength delaminate at the fiber–coating interface. Due to a higher shear strength the maximum bonding strength of Twaron is about 7 $N/mm^2$. None of the oxide ceramic coatings reach this limit, but by the use of a bond coat, the bonding strength on Twaron is increased. Especially the $Al_2O_3$-AlSi coating shows a high bonding strength. All coatings deposited on the Twaron fabrics delaminated at the fiber–coating interface.

In order to ensure identical spray parameters the investigated fabrics were coated together within one coating process. The differences in the mechanical properties of the coated fabrics are obvious, concerning abrasive behavior against metal counterparts as well as bonding strength of the thermally sprayed coatings. This might also be due to the differences in the process temperatures since the fabrics have different thermophysical properties, which influence the wetting and bonding behavior of the applied coatings. For example, due to the high thermal load a certain fiber damage of the Twaron fibers during the coating process might cause a decreased bonding strength of the oxide ceramic coatings on the fiber fabric.

CONCLUSIONS

The approach to combine highly refractory oxide ceramic coatings with high modulus lightweight fiber fabrics has been successfully demonstrated. Atmospheric plasma spraying with well defined parameter sets and simultaneous cooling is a suitable process for the coating of oxide ceramics on top of fiber woven fabrics for ballistic protection. Even though the $TiO_2$- and $Al_2O_3$- coatings have melting points above 1800° and. 2000°C respectively, no significant polymer fiber damage has been seen. The adherent coatings remain flexible and reach a hardness up to 1240 HV 0,05. The bonding strength is sufficient and mainly limited by the maximum shear strength of the fibers. The adhesion of the coatings and the high cycle flexibility can be improved by using metallurgical bond coats. So far the best results have been reached with an $Al_2O_3$- coating on a Twaron[©] fabric with a AlSi bond coat. It has the highest microhardness, best ability to blunt a metal counterpart and the highest bonding strength. Further efforts will focus on the optimization of the interface between oxide ceramic coating and fiber fabrics by tailoring the cooling process during thermal spraying as well as by deposition of metallurgical thin films as bond coats.

ACKNOWLEDGMENT

The authors would like to thank Mrs Katrin Keck (metallography) and Mr Chuanfei Li (plasma spraying) for their support and Mr Scherer for the helpful discussions.

REFERENCES

[1] C. L. Segal, "High- performance organic fibers, fabrics and composites for soft and hard armor applications," *23rd International SAMPE Conference*, USA, 651-660 (1991)

[2] N. Laetsch, "The use of Kevlar para- aramide fiber in ballistic protection garments," Du Pont de Nemours International SA, *Technical Textiles International*, Elsevier Science Ltd Oxford England, 26, (1992)

[3] J. van Dingenan and A. Verlinde, "Nonwovens and fabrics in ballistic protection," *Technical Textiles International*, Elsevier Science Ltd Oxford England, 10-13, (1996)

[4] P. G. Riewald, H. H. Yang, W. F. Shaughnessy, "Lightweight Helmet From A New Aramide Fiber," Du Pont Fibers, Wilmington, (1991)

[5] European Aeronautic Defense and Space Company EADS *www.army-technology.com* (Dec. 2000)

[6] J. Weatherall, M. Rappaport, J. Morton, "Outlook for advanced armor materials," National SAMPE Technical Conference *Advanced Materials*: Looking Ahead of the 21st Century, 1070-1077, USA, (1990)

[7] J-P. Charles, D. Guedra- Degeorges, "Impact Damage Tolerance of Helicopter Sandwich Structures," Aerospatiale, France (1999)

[8] Security Sicherheitstechnik GmbH *www.policeshop.de* (Dec. 2000)

[9] "Product data sheet Twaron", Twaro Products, D- 42097 Wuppertal, Kasinostraße 19-21, (1995)

[10] C. Friedrich, G. Berg, E. Broszeit, C. Berger: "Datensammlung zu Hartstoffeigenschaften," *Materialwissenschaft und Werkstofftechnik*, Vol. 28, No. 2, (1997)

[11] L. Pawlowski, "The science and engineering of thermal spray coatings," John Wiley and sons, Chichester (1995)

[12] "Das Verfahrensspektrum beim thermischen Spritzen," Linde AG, Werksgruppe technische Gase, Höffriegelskreuth (1990)

[13] G. Pursche, "Oberflächenschutz vor Verschleiß, Verlag Technik, Berlin (1990)

[14] M. Buchmann, R. Gadow, D. Scherer, "Mechanical and tribological Characterization of $TiO_2$ based multilayer coatings on light metals," The 25th Annual

International Conference on Advanced Ceramics and Composites, Cocoa Beach, 21-26, (2001); *in print*

[15] A. Skopp, M. Woydt, "Ceramic and Ceramic Composite Materials with Improved Friction and Wear Properties," *Tribology Transactions*, 48th Annual Meeting in Calgary, Alberta, Canada, Germany, 17-20, (1993)

[16] M. Woydt, "Werkstoffkonzepte für den Trockenlauf", *Tribologie und Schmierstofftechnik*, (1997)

# Laminated Object Manufacturing

# THE EFFECT OF SHEETSTOCK ON THE LAYERED MANUFACTURING OF A TELESCOPING CERAMIC ACTUATOR

Barry A. Bender and Roy J. Rayne
Code 6351
U.S. Naval Research Laboratory
Washington, DC 20375-5343

ABSTRACT

Layered manufacturing of a novel telescoping ceramic actuator was done using green ceramic tapes from two commercial sources. The characteristics of the starting sheetstock affected the lamination, laser cutting, and firing of the laminated object manufacturing (LOM) process.

INTRODUCTION

One solution to reduce the cost and lead-time for insertion of specialized functional ceramics is solid freeform fabrication (SFF). SFF generates dimensionally accurate prototypes from CAD files without the need for expensive tooling, dies, or molds. As a result, novel components in the green state can be manufactured in one operation. Fabrication of production components can be created for immediate 'form, fit, and function' performance evaluation. Rapid iterative design optimization is possible since modified tooling is not required. All of this leads to a time-compressed development cycle of the ceramic component.

One SFF manufacturing process that has been successful in building ceramic components is the layered manufacturing process developed by Griffin et al.[1] They modified the commercial LOM process developed by Helisys (Torrance, CA) for SFF of design concept models to use ceramic sheetstock instead of laminates of paper during the component build. The makeup of the ceramic sheetstock is critical to the success of the process. The ideal tape feedstock has to be strong enough to handle without distortion, compatible with laser cutting, and it has to be suitable for lamination via solvent welding while still possessing appropriate firing properties.[2] Often these requirements conflict with each other and tradeoffs are necessary. This paper reports the effects of two different lead zirconate titanate (PZT) tape systems on the LOM processing parameters during layered manufacturing of a novel telescopic actuator.[3]

EXPERIMENTAL PROCEDURE

The LOM Process

A commercial LOM system (2030 H, Helisys, Torrance, CA) was used to fabricate the actuators. Typically, fabrication is accomplished by feeding LOM paper (paper with a heat-sensitive adhesive backing) to the build and laminating it to the built-up stack using thermocompression via a heated roller. The thickness of the build is then measured. This information is sent to the LOM computer that calculates the 2-D slice for that particular thickness from the model's CAD file. The computer drives a 50 watt $CO_2$ laser to cut the perimeter of that cross section one layer deep. The unwanted material is laser-diced. This process is repeated layer by layer until the part is finished. The build is then removed from the machine and the waste material is removed (decubed) resulting in the fabricated 3-D part.

Actuator Fabrication

To fabricate the PZT actuator, the commercial LOM process was modified in several ways. The LOM paper was replaced by sheetstock of two different commercial green PZT tapes that were 0.2 mm thick. The tapes were cut into sheetstock (4 cm x 4 cm) and placed manually onto the build. Lamination via the heated roller led to poor lamination efficiency so solvent welding was used. Solvent welding was done by spraying each tile with a controlled amount of solvent. The tape was then bonded to the build using compression via a hand roller. Typically 75 to 80 layers of tape were used to build an actuator that was 2.5 cm in diameter and 1.2 cm in length. Decubing in hard-to-reach areas of the build was done after laser cutting of a laminated layer. The decubed actuator was then fired slowly to 550°C to burn out the binder and bisqued to 780°C in order to handle the part.[4] The actuator was then placed on setter sand and fired to 1245°C for 2 hours in an enclosed environment.

Processing Parameters Investigated

The effect of two commercial PZT tapes with different binder systems (designated as tape A and tape B) on the above process was investigated. Research was done on the lamination and laser cutting characteristics of both tapes. A study was done on each tape to determine what the best solvent to use in order to achieve good lamination efficiency via solvent welding. Laser cutting parameters (speed and power) were investigated for each tape. This was done by determining 3 different sets of laser cutting parameters in which the LOM laser barely overcut into the next layer of tape. These parameters are listed in Table I.

Innovative Processing and Synthesis of Ceramics

Last of all, problems that arose during firing of the built actuators were investigated. All microstructural characterization was done either by optical or scanning electron microscopy (SEM).

## RESULTS AND DISCUSSION
### Microstructure of the PZT Tapes

The tape A PZT sheetstock were tapes incorporating a binder system optimized for alumina tapes that had been specially formulated for their ease of lamination via solvent welding. The tapes consisted of 0.5 μm PZT particles in a proprietary binder system. The tapes were quite flexible, as additional plasticizer was needed to handle the high specific gravity of PZT. One side of the tape has a 2 μm thick layer of binder-rich material, which is typical for tapecast ceramics. The binder is not homogeneously distributed, as it was common to find pockets of binder with very little powder. The PZT powder is not evenly distributed either. As seen in a polished cross section of a fired actuator (Fig. 1A) porosity decreases significantly going from the top to the bottom of the tape. This same effect was also observed in single layers of the tape eliminating the lamination process as a possible cause for this problem. The tape lost 16.9% of its weight during binder burnout.

The tape B sheetstock were commercial PZT tapes designed for making multi-layer capacitors. The tapes consisted of 0.4 μm PZT particles in a proprietary binder system where the PZT volume loading was similar to tape A. The tape B binder system is different as it was designed to allow for lamination via thermocompression instead of solvent welding. The tapes are much stiffer than tape A, but they handle well. One side of the tape has a 2 μm thick layer of binder-rich material also. The binder was homogeneously distributed around the PZT particles. No density gradients were observed going from top to bottom (see Fig. 1B). The tape lost 9.7% of its weight during binder burnout.

### Lamination Parameters

Initially, lamination of the layers to the build was done via thermocompression by the heated steel roller of the LOMS. With tape A the sheetstock stuck to the heated roller due to the amount of plasticizer in the tape. With tape B the initial attempt of lamination via thermocompression was thought to be successful. However, upon decubing the build the actuator was distorted due to shearing from the pressure of the heated roller. Using less pressure did not solve the shearing problem and increasing the temperature of the roller to achieve better tackiness resulted in tape B sticking to the roller also.

Solvent welding had to be utilized to achieve reliable repetitive layer lamination. To optimize lamination efficiency the proper solvent has to be ascertained. The solvent must act as a tackifier enabling bonding between layers.

It also must have a relatively low vapor pressure so that it doesn't evaporate during spraying or before laminating. It must not attack aggressively the binder system of the tape or decubing of the waste material from the build will be difficult without distorting the part. At the same time it must soften the layer interface to allow good conformability between the layer and the build using very low lamination strain. Previous research using tape A indicated that a thin sprayed coating of propanol led to effective lamination (i.e.eradication of a layered structure).[5] However, on closer examination 100% lamination efficiency was not being achieved as lamination debonds as wide as 25 µm were observed scattered throughout the cross section of an actuator build. Fracture surfaces of a green build broken in liquid nitrogen show the variable presence of small clusters of elongated pores side by side (Fig. 2A) at the interface between layers of tape. These defects were linked to the hand rolling lamination process. Small bubbles were seen to form under the new layer of tape during the rolling process. These bubbles are pockets of excess solvent that could not be eliminated by hand rolling due to the suppleness of the tape.

Due to its different binder system, propanol was not an effective lamination assist solvent for lamination of tape B. Research led to the use of xylenes for the best balance between appropriate vapor pressure, tackiness, and softening of the tapes. Due to the stiffness of the tape bubbles were not observed during rolling of the sheetstock to the build. However, SEM characterization of fracture surfaces of a green actuator build indicated that defect-free lamination was not obtained. As seen in Fig. 2B a limited number of defects are seen throughout the fractured cross section. It is believed that these defects are the result of poor local particle packing at the interface.[6] This results because the xylenes don't soften the tape/build interface enough to produce intimate contact along 100% of the entire interface.

Laser Cutting Parameters

For optimum laser cutting of ceramic materials one would use high power and high speed.[7] However, due to safety considerations and cutting accuracy, commercial LOM systems use low power and low speed. This leads to cut edges that are jagged, as there is not enough power being used, which causes the resultant kerf to be slanted inwards. Also low power and low speed leads to the formation of a heat affect zone[2] (HAZ) in the cutting of green ceramic sheetstock. This is a result of heat being conducted away from the cutting zone and melting the organic components in the tape in a zone next to the cut leaving that area more brittle. To reduce these effects research was done to establish the best laser cutting parameters for each tape. The experimental results showed similar trends for both tapes. As expected, the greater the power the wider the kerf (see Fig. 3), and the greater the power the narrower the HAZ (see Table I).

Innovative Processing and Synthesis of Ceramics

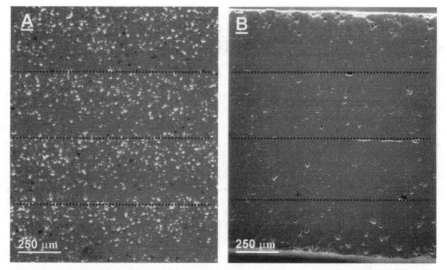

Fig. 1 SEM micrographs of the polished cross sections of fired layups made with (A) tape A sheetstock showing density gradients going from the top to the bottom of each layer of tape, while (B) shows no density gradients exist in the tape B sheetstock.

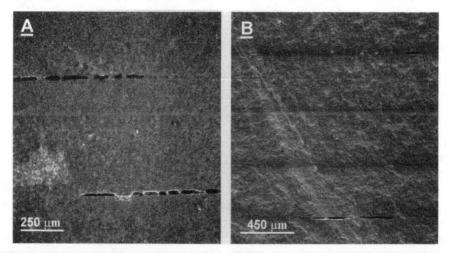

Fig. 2 SEM micrographs of the fracture surfaces of the green build layups made with (A) tape A sheetstock showing 25 micron wide defects due to entrapped bubbles of propanol, while (B) shows a limited number of processing defects found in the tape B sheetstock which were laminated together using xylenes.

However, due to the tapes having different binder systems their laser cutting behavior differed somewhat. As seen in Table I slower cutting leads to a larger HAZ. This is due to the fact that there is more time for some of the laser energy to be drawn away from cutting the kerf and into melting the plasticizer in the tape next to the cut. This can be seen in Fig. 3 where there were more small amorphous-looking balls present on the surface of the slower cut. These balls are the organic components of the tape that have boiled out. Due to the presence of a greater volume percent of binder in the tape A sheetstock, and therefore less porosity, heat conduction into the tape is greater. This leads to a larger HAZ (see Table I). It also leads to poorer kerf quality. This can be seen in Fig. 4 where the tape A cut is more V-shaped and that there is more damage done to the underlying tape. This also explains why less power is needed to cut tape B. This is verified by the power/speed ratios listed in Table I. For the tape B sheetstock heat conduction away from the cut does not become a significant factor until the slowest cutting speed where the power/speed ratio suddenly doubles. In case of the tape A sheetstock the power/speed ratio drops as expected with increasing speed but never plateaus indicating lateral heat conduction is still a factor even at the highest cutting speed.[7]

Table I. Effect of Ceramic Tape on LOM Laser Cutting Parameters

| Green PZT Tape Tested | Laser Cutting Power (%) | Laser Cutting Speed (cm/s) | Laser Power/ Laser Speed | HAZ Size (μm) |
|---|---|---|---|---|
| Tape A | 100 | 20 | 5.0 | 55 |
| Tape A | 81 | 10 | 8.1 | Not measured |
| Tape A | 64 | 5 | 12.8 | 125 |
| Tape B | 82 | 20 | 4.1 | 25 |
| Tape B | 52 | 10 | 5.2 | Not measured |
| Tape B | 46 | 5 | 9.2 | 90 |

Firing

After firing to 1245°C the tape B-based actuator showed better integrity than the tape A-based actuator. The tape A-based actuator had problems with deforming, cracking, and delaminating (see Fig. 5A). These problems were related to the greater amount of plasticizer used in its binder system. This was confirmed by firing a single layer of tape A sheetstock and observing the great amount of warpage it underwent. In addition to the amount of plasticizer there are two additional causes for the cracking and distortion. Inhomogeneous binder distribution leads to internal density distributions after binder burnout, which can lead to distortion and cracking during sintering.[8] Also the density gradients as seen Fig. 1A will lead to warpage problems and poor conformability between the new tape and the build, which can lead to the observed lamination problem.[9] The

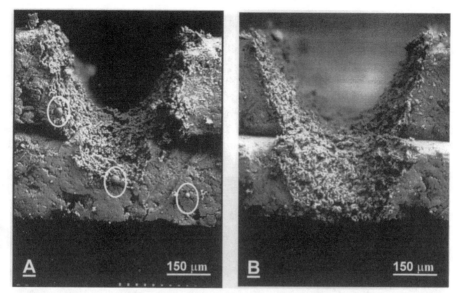

Fig. 3 SEM micrographs of a layer of tape B cut with the LOM laser at (A) 5 cm/s using 46% power and (B) 20 cm/s using 82% power. Note the presence of small amorphous-looking balls near the kerf in figure A.

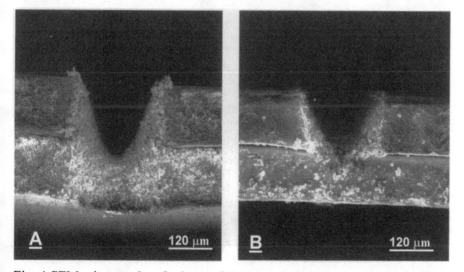

Fig. 4 SEM micrographs of a layer of (A) tape A and (B) tape B laser cut at the same speed showing differences in kerf quality.

lamination problem was also exacerbated by entrapped bubbles formed during rolling due to the suppleness of the tape. To overcome the deformation problem a sacrificial end cap was added to the tape A-based actuator (see Fig. 5B) resulting in a more robust actuator (Fig. 5C).

The tape B-based actuator did not need the addition of a sacrificial end cap. The actuator after firing showed good integrity (Fig. 5D). Its cut surfaces were smoother and didn't blister off like the tape A-based actuator. This was due to tape's binder system, which led to a smaller HAZ during LOM laser cutting. Microstructural characterization of the fired fracture surface did show that improvements in lamination efficiency are still needed.

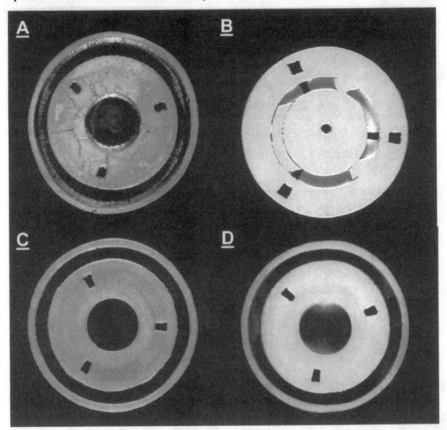

Fig. 5 Optical micrographs of the bottom view of fired tape A-based actuators (A-C) and (D) a fired tape B-based actuator (all actuators are 2. 5 cm in diameter).

CONCLUSION

The attributes of the starting green sheetstock greatly influence the success of the modified LOM process. Though both binder systems were amenable to solvent welding they required different lamination assist solutions. Stiffer tapes led to better expression of the lamination solvent and elimination of bubble pockets of liquid. Sheetstock with less volume percent organic constituents lead to higher quality LOM laser cutting. Tapes with less plasticizer and homogeneous distribution of binder and powder resulted in less deformation and cracking during firing. Therefore, to make specialized functional ceramics via layered manufacturing great care must be taken in designing the green ceramic tape to ensure component reproducibility and performance.

REFERENCES

[1]C. Griffin, J. Daufenbach, S. McMillin, "Desktop Manufacturing: LOM vs Pressing," *Amer. Ceram. Bull.*, **73** [8] 109-13 (1994).

[2]J.D. Cawley and Z. Liu, "Applying Tape Casting to Layered Manufacturing Processes," *Ceram. Ind.*, **128** [3] 42-48 (1998).

[3]C. Cm. Wu, D. Lewis, M. Kahn, and M. Chase, "High Authority, Telescoping Actuators," *Proc. SPIE*, **3674** 212-19 (1999).

[4]A. Bandyopadhyay, R. K. Panda, V. F. Jana, M. K. Agarwala, S. C. Danforth, and A. Safari, "Processing of Piezocomposites by Fused Deposition Technique," *J. Am. Ceram. Soc.*, **80** [6] 1366-72 (1977).

[5]B. A. Bender, R. J. Rayne, and C. Cm. Wu, "Solid Freeform Fabrication of a Telescoping Actuator Via Laminated Object Manufacturing," *Ceram. Eng. Soc. Proc.*, 21 [4] 143-50 (2000).

[6]J. Jean, and H. Wang, "Organic Distributions in Dried Alumina Green Tape," *J. Am. Ceram. Soc.*, **84** [2] 267-72 (2001).

[7]J. D. Cawley, Z. Liu, J. Mou and A. H. Heuer, "Materials Issues in Laminated Object Manufacturing of Powder-Based Systems," *Proc. Solid Freeform Fab. Symp.*, Univ. of Texas at Austin, 503-09 (1998).

[8]P. K. Lu, J. J. Lannutti, "Effect of Density Gradients on Dimensional Tolerance During Binder Removal," *J. Am. Ceram. Soc.*, **83** [10] 2536-42 (2000).

[9]N. Suppakarn, H. Ishida, and J. D. Cawley, "Roles of Poly(propylene glycol) during Solvent-Based Lamination of Ceramic Green Tapes," *J. Am. Ceram. Soc.*, **84** [2] 289-94 (2001).

# Electronic and Magnetic Materials

# BINDER REMOVAL BY SUPERCRITICAL EXTRACTION FROM BaTiO₃-Pt MULTILAYER CERAMIC CAPACITORS

Stephen J. Lombardo and
Rajesh V. Shende
Department of Chemical Engineering
University of Missouri
Columbia, MO 65211, USA

Daniel S. Krueger
Honeywell
Federal Manufacturing
& Technologies, LLC
Kansas City, MO 64141, USA

## ABSTRACT

As multilayer ceramic capacitors (MLCs) become larger for increased charge storage, the removal of binder by thermal degradation becomes more difficult. An alternative strategy is to remove the binder by supercritical extraction (SCE). In this work, we present the results for the degree to which poly(vinyl butyral) (PVB) and dioctyl phthalate (DOP) can be removed from MLCs. Whereas supercritical carbon dioxide can easily extract the DOP plasticizer, removal of the PVB is more difficult. Alternate strategies for PVB removal are discussed, and we assess the degree to which the permeability of the body is enhanced by using supercritical extraction to increase the pore space in the body.

## INTRODUCTION

Removal of the binder without degradation of the ceramic component is one of the problems encountered in processing ceramic materials for various applications [1]. The common method of binder removal is by thermal degradation of the polymers [2,3]. This process is accompanied by several deficiencies, the major ones being:
- The generation of degradation products leads to an increase of pressure, which can cause bloating, fracture or delamination of the component [4,5].
- Some carbon may be left in the body [6] that may modify the electrical properties and adversely influence the sintering behavior.

An alternative method to thermal degradation of polymeric binders is supercritical extraction [7-10]. In this approach, a gas such as $CO_2$ is taken to conditions above its critical temperature of 304.2 K and critical pressure of 7.38 MPa. In this state, the supercritical $CO_2$ possesses an interesting mixture of properties being neither truly liquid or gas. The property of interest here is that supercritical $CO_2$ can be used to dissolve polymeric materials and thus can remove binder from green ceramic components without generating appreciable stress within the body.

In most applications, supercritical $CO_2$ has been used alone as the extraction medium. It is well known, however, that addition of other polar and nonpolar liquids, called entrainers, may enhance the solubility of a polymer in supercritical fluids. This is especially important for high molecular weight polymers that typically have lower solubility than low molecular weight components [11]. Another advantage of supercritical fluid extraction is that the solubility of the polymer is a function of the solvent density, which can be changed by varying both pressure and temperature. In this project, we have explored the extraction of binder using supercritical carbon dioxide for $PVB+DOP+BaTiO_3$ and $PVB+DOP+BaTiO_3+Pt$ multilayer ceramic capacitors.

## EXPERIMENTAL

The $BaTiO_3$ green bodies used for supercritical extraction were found to have 10.89 w% of polymeric binder. The binder, after removal of the volatile solvents, is a mixture of PVB (55 w%) and DOP (45 w%). The MLC components were in the shape of rectangular parallelepipeds with nominal side lengths of 1.5 cm by 1.4 cm by 0.25 cm height.

The binder removal experiments were conducted in a 500-ml Parr high-pressure vessel with provisions for temperature and pressure measurement, and the apparatus is illustrated in Fig. 1. Experiments were carried out at 35, 55, 75, and 95°C in the pressure range of 10 to 40 MPa at 10 MPa intervals. To maintain the temperature, the vessel was kept in a constant temperature bath.

In the supercritical extraction experiments, the degree of binder removal is reported in terms of two time variables: residence time and cycle time. The residence time indicates the total period of time a sample was in contact with a single charge of supercritical $CO_2$. For example, a 5-hour residence time means that the sample was in contact with one charge of supercritical $CO_2$ for 5 hours. A cycle time of 1-hour indicates that the sample was contacted with supercritical $CO_2$ for one hour and then the gas was released. A cycle time of 5-hours thus denotes that five 1-hour $CO_2$ charge-discharge cycles were used.

## RESULTS AND DISCUSSION

The initial binder removal experiments in supercritical $CO_2$ were conducted with $BaTiO_3$ MLC substrates containing no platinum metal electrode layers. Figure 2 indicates that the amount of binder removal first increases as a function of residence time and then reaches a plateau after three hours. The trends in Fig. 2 also indicate that increasing the temperature and pressure leads to more removal of the binder. A maximum of 42.8% binder removal by weight is achieved for a

Innovative Processing and Synthesis of Ceramics

3-hour residence time at 95°C and 40 MPa. The plateau beyond two hours indicates that the $CO_2$ is saturated with binder and that further time would not lead to more removal. At temperatures near and above 95°C, thermal degradation of the binder becomes appreciable and damage to the sample occurs.

Experiments were also conducted as a function of cycle time, and this data is displayed in Fig. 3 along with binder removal data as a function of residence time. For times of one hour, the amount of binder removed is equivalent by the two methods. This is expected, because a 1-hour cycle time is equivalent to a 1-hour residence time with both denoting a single exposure to $CO_2$. Beyond one hour, however, the amount of binder removed increases more with cycle time than with residence time. This is also expected, since fresh $CO_2$ is charged whenever the cycle time is more than one hour, and thus the supercritical fluid does not remain saturated with binder. For exposures beyond three hours, increasing the cycle time does not lead to further removal of binder. Maximum binder removal of 52% was observed for a three hour cycle time as compared to 40% removal for a 3-hour residence time at 75°C and 40 MPa.

Figure 4 represents the degree of binder removal as a function of pressure for cycle times of three hours over a range of operating conditions. For the two highest temperatures, the degree of binder removal increases rapidly at pressures from 10 to 30 MPa and then is seen to reach a plateau between 30-40 MPa, thus indicating that further increases in pressure would not drastically improve the amount of binder extracted. Based on these results, the maximum degree of binder removal in supercritical $CO_2$ occurs at 40 MPa and 95°C when a 3-hour cycle time is used. At these conditions, however, damage to the samples was occasionally observed because the temperature is too close to the point at which thermal degradation becomes appreciable. Operating at 75°C and 40 MPa of $CO_2$ pressure, however, only leads to a few percent reduction in the degree of binder removal while maintaining the integrity of the samples, as evidenced by visual and light microscope evaluation of the samples at 5-20X.

Extraction experiments in supercritical $CO_2$ were also conducted with the pure PVB polymer. For this case, approximately 1.6% of the PVB is removed at 95°C and 40 MPa, which is much less than the 40-50% total binder removal seen in Figs. 2-4. For experiments conducted with DOP at the same conditions, 100% of this component could be extracted. Since the binder is a mixture of PVB (55 w%) and DOP (45 w%), the component of the binder being removed is the low molecular weight dioctyl phthalate. This interpretation is reasonable, because the extraction of polymers becomes more difficult as the molecular weight increases [12].

To enhance the degree of binder removal, entrainers such as 2-propanol, methyl isobutyl ketone (MIBK), and n-hexane were evaluated. Five milliliters of a single solvent were charged into the vessel for cycle times of 1, 3, and 5 hours.

As seen in Fig. 5, the effect of propanol and MIBK as entrainers on the degree of binder removal is not appreciable. The use of n-hexane as an entrainer also did not lead to an improvement in the degree of binder extraction.

From the results presented above, we can observe that at 95°C and 40 MPa $CO_2$ pressure for three 1-hour cycles, the low molecular weight DOP has been almost completely extracted from the MLCs. At these conditions, however, the PVB has not been removed by the supercritical $CO_2$. To enhance the removal of the PVB, one possibility is to add oxygen at 95°C so that some thermal oxidative degradation occurs, and the supercritical fluid can then extract the lower molecular weight degradation products. This approach was used at 95°C and 40 MPa total pressure with an oxygen partial pressure of 7 MPa for 1 hour followed by two 1-hour cycles in pure $CO_2$ at 40 MPa. With this approach, only 56 weight% of the binder could be extracted, which is only 1-2% more than when oxygen was not present.

To model the time to remove binder from the porous ceramic bodies, we treat the extraction process first to be the dissolution of the binder followed by diffusion out of the body [9,10]. For a 3-dimensional body of side lengths x=2a, y=2b, and z=2c, unsteady state binder removal by a diffusive process can be described by [9]:

$$\frac{\partial C}{\partial t} = D\left(\frac{\partial^2 C}{\partial x^2} + \frac{\partial^2 C}{\partial y^2} + \frac{\partial^2 C}{\partial z^2}\right) \qquad 1$$

with boundary conditions that at time $t=0$, the concentration $C=C_o$ everywhere in the body and at the body edges, x=±a, y=±b, z=±c, $C=0$ for all times. The latter condition was not satisfied precisely in our experiments because we operated in a semi-batch mode.

The solution to Eq. 1 with the prescribed boundary conditions for the average concentration in the solid is:

$$\frac{\overline{C}}{C_o} = \frac{512}{\pi^6} \sum_{i=0}^{\infty} \sum_{j=0}^{\infty} \sum_{k=0}^{\infty} \frac{1}{[(2i+1)(2j+1)(2k+1)]^2} \exp(-\alpha_{ijk}t) \qquad 2$$

where

$$\alpha_{ijk} = \frac{D\pi^2}{4}\left[\left(\frac{2i+1}{a}\right)^2 + \left(\frac{2j+1}{b}\right)^2 + \left(\frac{2k+1}{c}\right)^2\right] \qquad 3$$

For the experiments conducted here, a=0.74 cm, b=0.69 cm, and c=0.13 cm. Figure 6 illustrates the model prediction for the time dependence of binder removal with a diffusivity, $D=1\times10^{-10}$ m²/s, and it can be seen that the model provides a reasonable description of the binder removal process. Equation 2 can thus be used to predict the time for binder removal for bodies of different dimensions.

A summary of the degree of binder removal at different conditions is given in Table I. The best conditions for supercritical extraction are 75-95°C and 40 MPa $CO_2$ pressure for a 3-hour cycle time. Under these conditions, 50-55% of the binder was removed from MLC substrates. Use of entrainers or a partial pressure of oxygen was not seen to enhance significantly the degree of binder removal.

Table I. Percent binder removal with supercritical $CO_2$ extraction under various conditions for a 3-hour cycle time.

| Composition | Entrainer | T (°C) | $P_{Total}$ (MPa) | % Binder Removal |
|---|---|---|---|---|
| Pure PVB | -------- | 95 | 40 | 1.6 |
| Pure DOP | -------- | 95 | 40 | 100.0 |
| PVB+DOP | -------- | 95 | 40 | 52.3 |
| PVB+DOP+BaTiO$_3$ | -------- | 95 | 40 | 55.7 |
| PVB+DOP+BaTiO$_3$ | 2-propanol (5 ml) | 95 | 40 | 56.0 |
| PVB+DOP+BaTiO$_3$ | 2-propanol+MIBK+ n-hexane (5ml) | 95 | 40 | 56.5 |
| PVB+DOP+BaTiO$_3$ | 2-propanol+MIBK+ n-hexane (5ml) | 95 | 7 (O$_2$)+33 (CO$_2$)* | 56.9 |
| PVB+DOP+BaTiO$_3$ | -------- | 75 | 40 | 52.9 |
| PVB+DOP+BaTiO$_3$ | 2-propanol (5 ml) | 75 | 40 | 52.9 |
| PVB+DOP+BaTiO$_3$ | 2-propanol (5 ml) | 75 | 7 (O$_2$)+33 (CO$_2$)* | 53.0 |
| PVB+DOP+BaTiO$_3$ | MIBK (5 ml) | 75 | 40 | 51.7 |
| PVB+DOP+BaTiO$_3$ | n-hexane (5 ml) | 75 | 40 | 52.6 |
| PVB+DOP+BaTiO$_3$ | 2-propanol+MIBK+ n-hexane (5ml) | 75 | 40 | 51.9 |

*Conditions used for oxidation: 1 hour at indicated temperature, pressure, and composition followed by exposure to $CO_2$ (40 MPa) for two 1-hour cycles.

In spite of the low degree of removal of the pure PVB, the extraction of the DOP increases the porosity by a factor of two in the samples. This additional porosity will greatly facilitate the transport of decomposition products out of the body when a sample first subjected to supercritical extraction is then heated during a separate thermal treatment to remove the remaining polymer. This can be assessed by calculating the permeability of a body undergoing SCE before thermal binder removal and for a body undergoing thermal binder removal only. The permeability, κ, can be represented by the Kozeny-Carmen form as

$$\kappa = \frac{\varepsilon^3}{k(1-\varepsilon)^2 S^2} \qquad\qquad 4$$

where $\varepsilon$ is the porosity, $k$ is factor to account for tortuosity and constrictions in the pores, and $S$ is the surface area per unit volume. For the samples evaluated here, the solids loading is 50% by volume. When SCE is not performed, the bodies have an initial porosity of 0.20 whereas when SCE has been performed and 50% of the binder has been removed, the bodies have an initial porosity of 0.35. From Eq. 4, the ratio of permeabilities for these two cases is $\kappa_{SCE}/\kappa_{Thermal}=10$, and we thus see that the permeability is increased by a factor of ten over the body which has not had binder removed by SCE. This will also lead to a much lower buildup of pressure during the subsequent thermal binder removal step.

## CONCLUSIONS

Supercritical extraction with carbon dioxide has been evaluated to remove binder from $BaTiO_3$ multilayer ceramic capacitors. At conditions of 75-95°C and 40 MPa of $CO_2$, approximately 50-55% of the binder could be removed, and this was attributed mainly to the dioctyl phthalate component. The use of 2-propanol, MIBK, and n-hexane as entrainers did not improve the amount of binder extracted. Extraction in supercritical carbon dioxide with oxygen present was also not seen to enhance binder removal. Although not all of the binder could be removed by SCE, the extraction of 50% of the binder leads to an increase in the porosity of the body that translates to a factor of 10 increase in the permeability. The more open pore structure resulting from supercritical extraction of the binder is thus expected to facilitate the removal of the remaining polymer during the subsequent thermal removal step.

## ACKNOWLEDGEMENT

This project was funded by Honeywell, FM&T, LLC which is operated for the United States Department of Energy, National Nuclear Security Agency, under contract No. DE-AC04-01AL66850.

## REFERENCES

[1]T. Chartier, M. Ferrato and J.F. Baumard, "Supercritical Debinding of Injection Molded Ceramics, " *J. Am. Ceram. Soc.* **78** [7], 1787-92 (1995).

[2]I.E. Pinmill, M.J. Edirisinghe and M.J. Bevis, "Development of Temperature Heating Rate Diagrams for the Pyrolytic Removal of Binder Used for Powder Injection Molding, " *J. Mat. Sci.* **27** [16], 4381-88 (1992).

[3]F.F. Lange, B.I. Davis and E.J. Write, "Processing Related Fracture Origins: IV, Elimination of Voids Produced by Organic Inclusions," *J. Am. Ceram. Soc.* **69** [1], 66-9 (1986).

[4]S.A. Matar, M.J. Edirisinghe, J.R.G. Evans, and E.H. Twizell, "Diffusion of Degradation Products in Ceramic Moldings during Thermal Pyrolysis: Effect of Geometry, " *J. Am. Ceram. Soc.* **79** [3], 749-755 (1996).

[5]L.C.-K. Liau, B. Peters, D.S. Krueger, A. Gordon, D.S. Viswanath, and S.J. Lombardo, "The Role of Length Scale on Pressure Increase and Yield of PVB-$BaTiO_3$-Pt Multi-layer Ceramic Capacitors During Binder Burnout," *J. Am. Ceram. Soc.*, **83** [11], 2645-53 (2000).

[6]C. Dong and H.K. Bowen, "Hot Stage Study of Bubble Formation During Binder Burnout," *J. Am. Ceram. Soc.* **72** [6], 1082-87 (1989).

[7]S.Nakajima, S. Yasuhara, and M. Ishihara, "Method of Removing Binder Material from a Shaped Ceramic Preform by Extracting with Supercritical Fluid," US Patent No. 4,731,208, March 15, 1988.

[8]T.Miyashita, Y. Ueno, H. Nishio, and S. Kubodera, "Method for Removing the Dispersion Medium from a Molded Pulverulent Material," US Patent No. 4,737,332, April 12, 1988.

[9]E. Nishikawa, N. Wakao and N. Nakashima, "Binder Removal from Ceramic Green Body in the Environment of Supercritical Carbon Dioxide with and without Entrainers," *J. Supercrit. Fluids* **4** [4], 265-69 (1991).

[10]T. Chartier, E. Delhomme and J. Baumard, "Mechanisms of Binder Removal Involved of Injection Molded Ceramics," *J. De Physique III* **7** [2], 291-02 (1997).

[11]M.A. McHugh and V.J. Krukonis, "*Supercritical Fluid Extraction: Principles and Practice*," Butterworths Publishers MA, 1986.

[12]T. Chartier, E. Delhomme, J.F. Baumard, P. Marteau, P. Subra and R. Tufeu, "Solubility in Supercritical Carbon Dioxide, of Paraffin Waxes Used as Binders for Low-Pressure Injection Molding," *Ind. Eng. Chem. Res.* **38** [5], 1904-10 (1999).

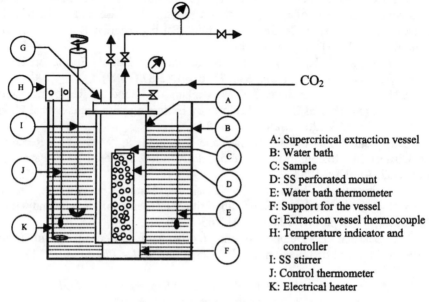

A: Supercritical extraction vessel
B: Water bath
C: Sample
D: SS perforated mount
E: Water bath thermometer
F: Support for the vessel
G: Extraction vessel thermocouple
H: Temperature indicator and controller
I: SS stirrer
J: Control thermometer
K: Electrical heater

**Figure 1.** Schematic of the supercritical extraction apparatus.

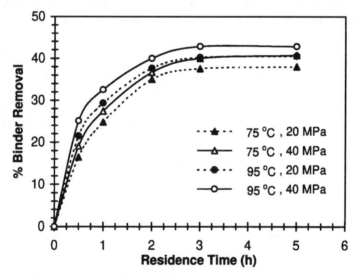

**Figure 2.** Degree of binder removal in supercritical $CO_2$ as a function of temperature, pressure, and residence time for $BaTiO_3$ MLCs.

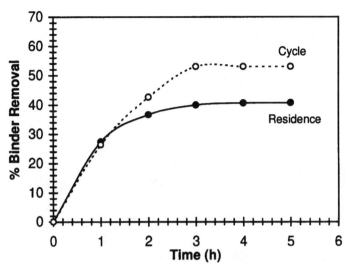

**Figure 3.** Comparison of the degree of binder removal from BaTiO$_3$ MLCs in supercritical CO$_2$ at 75°C and 40 MPa, as a function of residence and cycle time.

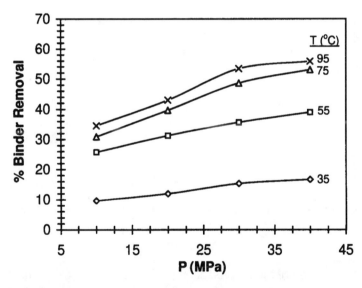

**Figure 4.** Degree of binder removal from BaTiO$_3$ MLCs in supercritical CO$_2$ as a function of temperature and pressure for a 3-hour cycle time.

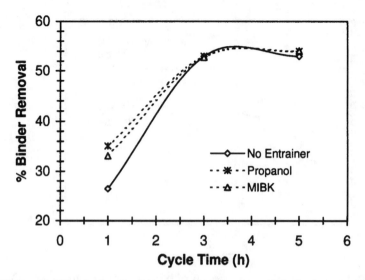

**Figure 5.** Effect of entrainers on the degree of binder removal with cycle time for BaTiO$_3$ MLCs in supercritical CO$_2$ at 75°C and 40 MPa.

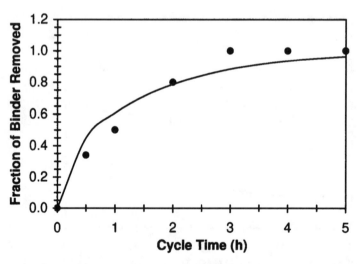

**Figure 6.** Comparison of model prediction (solid line) versus experimental results (filled circles) for binder removal from BaTiO$_3$ MLCs at 95°C and 40 MPa in supercritical CO$_2$. The best-fit value of the diffusivity is D=1 x 10$^{-10}$ m$^2$/s.

# PROCESSING AND CHARACTERIZATION OF PZT AND PLZT THICK FILMS

Oliver Huang, Amit Bandyopadhyay and Susmita Bose
School of Mechanical and Materials Engineering
Washington State University
Pullman, WA 99164-2920

ABSTRACT
Ferroelectric thin and thick films are being considered for numerous applications including micro-electromechanical systems (MEMS), transducers and actuators. Processing of ferroelectric films in the thickness range of 1 to 100 microns have already been attempted by various research groups, but failed in most of the cases due to extensive cracking problems during firing. For the past two years, we have worked on this problem and developed a process to fabricate thick films of lead zirconate titanate (PZT) and lead lanthanum zirconate titanate (PLZT) ceramics that were crack free, low in porosity with good ferroelectric properties. In our approach, sub-micron size powders of PZT and PLZT were synthesized via gelation of the precursor solutions. PZT and PLZT powders were then mixed with the same precursor solutions and ball-milled. Thick films of PZT and PLZT in the range of 5 to 20 μm were deposited via spin-coating technique on platinized Si substrates. The films were sintered at temperatures between 775 to 800 °C. Film thickness and surface roughness were characterized using SEM and profilometer. The surface and cross-sectional areas of the films appear to be crack free and homogeneous. Ferroelectric properties of some of these films are reported here.

INTRODUCTION
Processing of PZT and PLZT thin and thick films has become an increasingly popular field of study due to their applications in MEMS devices as well as transducers, sensors, and actuators [1-6]. Thick films offer advantages to generate larger actuating forces compared to thin films. However, the so-called "thick" film is only a relative term. The film thickness can be varied anywhere between 1 to 100 μm depending on the application and deposition technique.

Some approaches that have been recently studied include screen-printing, plasma spraying, sputtering and sol-gel [7-9]. The latter technique is of special interest to us because of its low cost of preparation and excellent stoichiometry control. Sol-gel deposition technique also offers the capability to lay down thick layers between 0.1 to 100 μm, a thickness range that is difficult to achieve by other deposition techniques. Several research groups have attempted to fabricate PZT and PLZT based thick films via sol-gel, however, problems such as cracks, low yields and poor fatigue properties occurred.

In this work, high quality PZT and PLZT thick films have been fabricated and characterized for their physical and electrical properties. The fabrication process entails utilizing a sol-gel method of dispersing powdered PZT or PLZT in a precursor solution to form homogeneous slurry. The slurry was then spin coated onto the platinized silicon substrates and sintered to achieve the final film. The processing part has two important areas including powder processing and film fabrication. The characterizations included phase identification of powders, slurry viscosity, solids loading, film thickness and roughness, ferroelectric fatigue and the influence of electric field on polarization.

EXPERIMENTAL PROCEDURES

*Processing of PZT and PLZT powders:* The powder processing work started with the preparation of the stock solution (sol) of the elements of the desired compounds, polymerization of the solution to form the gel and then drying and calcining the gel to form a final PZT or PLZT powder. The starting materials for $Pb_{1.1}(Zr_{0.52}Ti_{0.48})O_3$ and $(Pb_{1.1x}La_{1-x})(Zr_{0.52}Ti_{0.48})O_3$ powders were lead acetate trihydrate $(Pb(CH_3COO)_2.3H_2O)$, zirconium propoxide $(Zr(OCH_2CH_2CH_3)_4)$, titanium iso-propoxide $(C_{12}H_{28}O_4Ti)$ and lanthanum 2-methoxyethoxide. The lead acetate trihydrate was first dissolved in 2-methoxyethanol $(CH_3OCH_2CH_2OH)$ or 2-MOE solvent and distilled at 125°C for 2 to 3 hours to remove water. For PZT, the anhydrous solution was allowed to cool to room temperature before adding suitable amount of zirconium propoxide. The solution was then refluxed at 120 °C for 2 to 3 hours to promote adequate mixing of the lead and zirconium molecules. The solution was once again allowed to cool to room temperature before adding titanium iso-propoxide and was refluxed 120 °C for 2 to 3 hours. In the case of PLZT, lanthanum 2-methoxyethoxide was added to the anhydrous lead acetate solution prior to the addition of zirconium and titanium. Finally the solvent was distilled off to get the precursor sol of desired molarity. The pH of the precursor solution was adjusted to ~11 by adding nitric acid and then hydrolyzed using suitable amount of water to allow gelation to take place. The gel was dried at 110°C, crushed into powders, ball-milled, and finally calcined at 600°C for PZT and 800°C for PLZT to obtain the final PZT and PLZT powder. The final particle sizes were less than 1 μm (based on SEM micrographs).

Innovative Processing and Synthesis of Ceramics

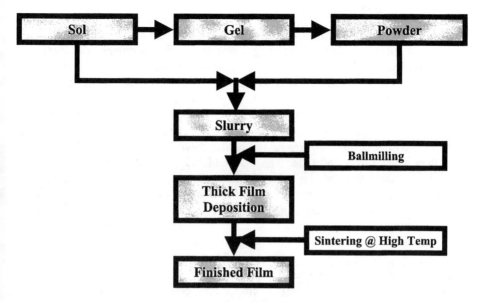

**Figure 1:** Process flow chart for PZT/PLZT powder and thick film

***Fabrication of thick films:*** **Figure 1** shows a process flow chart for the fabrication of ceramic powder and thick film using the same powder and the precursor sol. Thick film processing has three major steps including slurry preparation, spin-coating and sintering. Process optimization was carried out in all three steps to produce thick films with better quality having a higher yield and uniformity in terms of ferroelectric properties.

*Slurry Preparation:* The thick film slurry was prepared by mixing powders with their respective precursor solutions. Proper amounts of dispersant (Darvan-7) was added to reduce the viscosity and increase the solids loading. The slurry viscosity was measured using a Brookefield viscometer to optimize the slurry composition. The final slurry should have a high solids loading and yet not be so viscous that it cannot be uniformly deposited onto a platinized silicon substrate via spin coating. The final solids loadings were 38g powder and 12 g sol for PZT and 28 g powder and 12 g sol for PLZT. At the shear rate of 93 s⁻¹, slurry viscosities were 195 and 255 cp, respectively. To ensure slurry homogeneity with sub-micron-size powder particles within the slurry, the slurry was subjected to ball-milling for 2 to 4 days.

*Thick Film Deposition via Spin Coating:* Silicon wafers with (100) orientation with 300 nm thermally grown $SiO_2$, 20 nm of Ti and 200 nm of Pt layers deposited on top was chosen to be the substrate for thick film deposition. The platinum was used as the bottom electrode layer. The spin rotation and acceleration of the spin coater was controlled. The thickness of the film across the substrate was optimized as a function of spin rate and acceleration with single layer PZT films. With the desired spin rate and acceleration set, the slurry was placed on the substrate and spun for 35 seconds. Two spin coating techniques were used to observe their effects on film uniformity. The first technique, or the stationary method, allowed the slurry to be placed on a wafer prior to the spinning process. In the second technique, slurry was dispensed onto a wafer already in rotating motion, known as the rotational method. The rotational method produced significantly better quality films. After laying down the first layer of thick film, it was then pyrolyzed at 150°C for 5 minutes and 350°C for 10 minutes on hot plates to drive off the organics. To achieve thicker films with multiple layers, number of layers can be spun and the heat treatment process needs to be repeated after every layer. A final sintering is necessary after all the layers are spin-coated on to the substrate.

*Sintering of Thick Films*: The PZT and PLZT thick film samples were sintered in a muffle furnace at temperature for 775 to 800°C between 10 to 15 minutes. It was found that a higher sintering temperature potentially damage the bottom electrode. Various combinations of sintering time and temperature were chosen for sintering. The samples were characterized and compared to determine the best combination of time and temperature based on their ferroelectric properties. After sintering, the thickness for one layer sample was between 8 to 10 μm, depending on the solids loading in the slurry.

CHARACTERIZATION OF THICK FILMS
High temperature x-ray diffraction (HTXRD) was studied to understand the phase evaluation in the amorphous powders during calcining. The HTXRD studies were performed using a fully automated Philips X-pert system. A setting of 35 kV and 35 mA was applied to the cobalt target, with a step size of 0.02° (2θ) and a count time of 0.5s per step.
*Microscopy*: The microstructures of the PZT and the PLZT thick films were studied by a JEOL scanning electron microscope (SEM). The thick film sample was first cut in half using a scriber along the center. This allowed us to observe the effect of centrifugal force on the thickness uniformity.
*Profilometry*: The surface roughness analysis was measured using a profilometer (SPN Technology, Inc). The roughness data was collected based on the vertical movement displacement of the stylus as it scanned the surface of the substrates.
*Ferroelectric Property Characterization*: In order to measure the ferroelectric properties of the thick film capacitors, it was necessary to have top electrodes. Approximately 150 nm of gold were sputtered onto the wafers to fabricate circular top electrodes that were 0.5 mm in radius. The ferroelectric properties were measured using a probe station attached with a RT-66A with high voltage interface (HVI) and a high voltage amplifier (HVA). Since the thickness of single

layer of PZT or PLZT film was over 1 micron, high voltage amplifier was used to supply higher voltage to observe the ferroelectric behaviors. The hysteresis loop, effective dielectric constant ($K_{eff}$), maximum polarization, remnant polarization, and the coercive field at voltages between 150 V to 600 V were measured. The same setup was used to measure the ferroelectric fatigue properties of the thick films. Fatiguing the films at 35 kHz frequency, we were able to observe whether the films could withstand up to $10^9$ cycles, and if so, how the films were degraded over time.

RESULTS AND DISCUSSIONS
*PZT and PLZT powders:* The HTXRD scan for the amorphous PZT powder is shown in **Figure 2**. The results show that the amorphous to crystalline phase transition starts at 525°C and is complete by 575 °C without formation of any pyrochlore phase. For the PLZT powder, the crystallization starts at 600 °C and is complete by 700 °C as shown in **Figure 3**. Based on these results, the calcination temperatures for PZT and PLZT was selected at 600 and 700 °C, respectively.

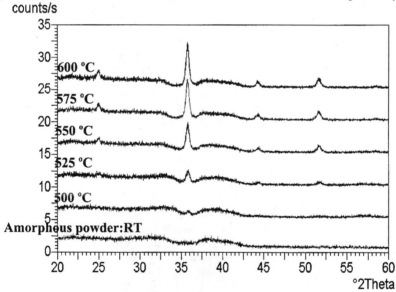

**Figure 2**: HTXRD plot of PZT powder showing the crystallization and perovskite phase formation

*PZT and PLZT Thick Films:* **Figure 4** shows the cross-sectional and top surface views of a single-layer PZT thick film. Film thickness was measured at increments of 0.5 mm. Ideally uniform thickness film is desirable, however, the results have shown that it is quite normal to have a thickness gradient from the center to the outer edge of the film. The ferroelectric property of the film depends heavily on

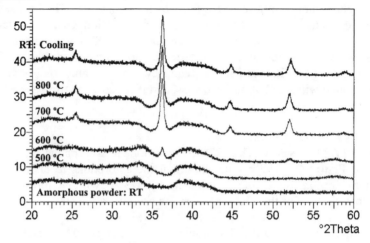

**Figure 3**: HTXRD plot of PLZT powder showing the crystallization and perovskite phase formation.

**Figure 4:** SEM images of (a) cross-sectional view of single-layer PZT thick film and (b) top surface at x10,000 magnification.

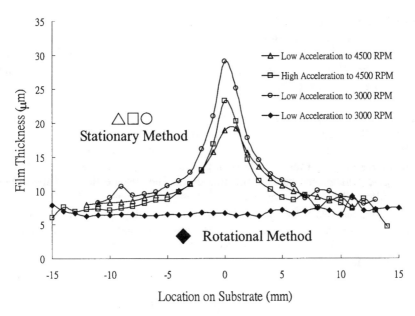

**Figure 5**: Single-layer PZT film thickness measurement at different location of the substrate deposited at different spinning rates and accelerations

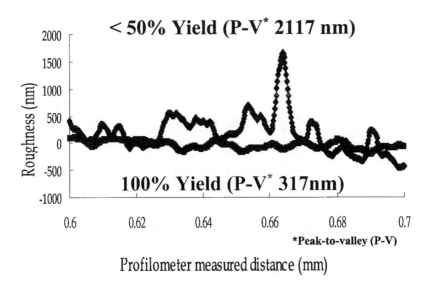

**Figure 6**: Profilometer measurement of surfaces of two single-layer thick film surface

the density of the film. Porosity was observed in the films due to incomplete sintering. Higher solids loading in the slurry improved the film properties as it reduced the residual porosity due to sintering. The high solids loading also minimizes the chances of delamination between the film and the platinum electrode due to increased shrinkage during sintering.

**Figure 5** shows a plot of film thickness as a function of the location on the substrate. For single layer PZT films deposited at different spinning rates and accelerations, a fairly uniform thickness film was formed except at the center of the substrate (~1 cm). **Figure 6** shows the surface roughness of single-layer thick film substrates measured using a profilometer. The surface that had the greater peak-to-valley distance showed a much lower percentage yield in comparison with the one that has significantly lower peak-to-valley distance. A smoother surface on the substrate can allow better top electrode (gold) adhesion, and subsequently contribute to improved contacts with the probes when conducting ferroelectric measurements.

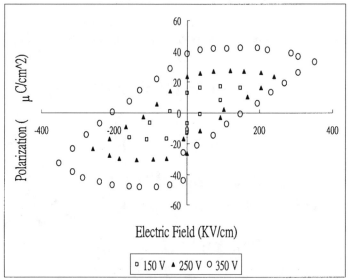

**Figure 7**: Hysteresis of single-layer 10 μm PZT film sintered at 775°C for 15 minutes.

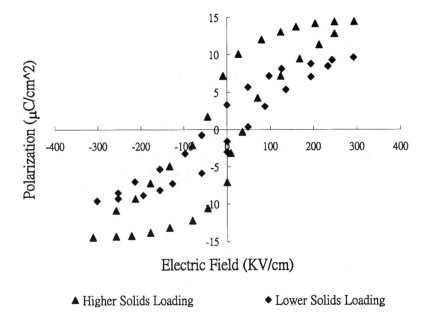

**Figure 8**: Hysteresis of single-layer PLZT (10 mm) showing higher polarization for slurry with higher solids loading.

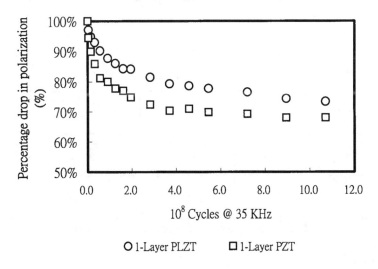

**Figure 9**: Fatigue result of a 10 μm single-layer PZT and PLZT thick films. Polarization degrades after $10^9$ cycles.

The hysteresis loops showed that the saturation and remnant polarizations are significantly higher for films prepared with slurry containing higher solids loading. **Figure 7** shows properties of a single layer PZT film as a function of applied voltage. **Figure 8** shows the influence of solids loading on the ferroelectric property improvements of these thick films. The initial solids loading for PZT was 71%, and that for PLZT was 67%, which was increased to 76% and 71%, respectively by adding more dispersant. **Table 1** shows percentage yields for both 1- and 2-layer PZT/PLZT thick films. Yielding was calculated based on consistent ferroelectric properties on 2.5 cm x 2.5 cm substrates that had 25 to 30 capacitors per substrate in average. As it was mentioned earlier, increasing solid loading in the slurry results in significantly better film qualities. Similar results were also obtained for PLZT films.

Research on the ferroelectric fatigue life of sol-gel derived PZT and PLZT thin films had shown the remnant polarization drops below 50% when films were fatigued past $10^8$ cycles. It is desirable to have fatigue greater than $10^{10}$ cycles for piezoelectric thin and thick films for devices that requires long-term applications. **Figure 9** shows the fatigue test result of single-layer PZT and PLZT thick films. By $10^9$ cycles the remnant polarization for PLZT has decreased from 10 $\mu C/cm^2$ to around 8 $\mu C/cm^2$, a drop of about 20%. The PZT thick film had lower initial polarization value, and also showed a larger drop (33%) after fatiguing for $10^9$ cycles.

| Sample | Spinning rpm | Sintering temperature | Soak time | Yield |
|--------|--------------|-----------------------|-----------|-------|
| 1 Layer PLZT | 4500 | 775°C | 10 Minutes | 95 % |
| 1 Layer PZT | 4500 | 775°C | 10 Minutes | 100 % |
| 2 Layer PZT | 4500 | 775°C | 10 Minutes | 100 % |

**Table 1**: Results of the yielding tests performed at 150V.

SUMMARY
The HTXRD tests showed that the PZT and PLZT powders can be calcined at 600 and 700 °C, respectively in phase pure perovskite phase. PZT and PLZT thick films in the range of 5 to 20 microns were successfully fabricated with high yield. These films are free of cracks and low in porosities. Solids loading of the slurry

plays a big role on the microstructure of the thick films and influences their electrical properties. Higher solids loading translates to better polarizations (both saturation and remnant) and fatigue properties with improved yields.

ACKNOWLEDGEMENTS
The authors would like to acknowledge the financial support from the Washington Technology Center, Seattle and ATL Ultrasound, Bothell, WA. Experimental helps from Mr. Todd Myers and Mr. Gabe Kuhn are also appreciated.

REFERENCES
1. D. A. Barrow, T.E. Petroff, M. Sayer, "Thick Ceramic Coatings Using a Sol Gel Based Ceramic-Ceramic 0-3 Composite", *Surface and Coatings Technology*, **76-77**, 113-18 (1995).
2. S. P. Beeby, A. Blackburn, N. M. White, "Processing of PZT Piezoelectric Thick Films on Silicon for Microelectromechanical Systems", *J. Micromech. Microeng.* **9**, 218-29 (1999).
3. X. N. Jiang, C. Sun, X. Zhang, B. Xu, Y. H. Ye, "Microstereolithography of Lead Zirconate Titanate Thick Film on Silicon Substrate", *Sensors and Actuators*, **87**, 72-77 (2000).
4. G. Yi, M, Sayer, "Sol-Gel Processing of Complex Oxide Films", *Ceramic Bulletin*, **70** [7] 1173-79 (1991).
5. R. Kurchania, S.J. Milne, "Characterization of Sol-Gel Pb($Zr_{0.53}Ti_{0.47}$)$O_3$ Films in the Thickness Range 0.25 – 10 $\mu$m", *Journal of Materials Research*, **14**[5] 1852-59 (1999).
6. R. Seveno, P. Limousin, D. Averty, J.-L. Chartier, R. Le Bihan, H.W. Gundel, "Preparation of Multi-Coating PZT Thick Films by Sol-Gel Method onto Stainless Steel Substrates", *Journal of the European Ceramic Society*, **20**, 2025-28 (2000).
7. W. Haessler, R. Thielsch, N. Mattern, "Structure and electrical properties of PZT thick films produced by plasma spraying", *Materials Letters* **24**, 387-391 (1995).
8. R. Maas, M. Koch, N.R. Harris, N.M. White, A.G.R. Evans, "Thick-film printing of PZT onto silicon", *Materials Letters* **31**, 109-112 (1997).
9. W. Zhu, K. Yao, Z. Zhang, "Design and fabrication of a piezo electric multilayer actuator by thick-film screen printing technology", *Sensors and Actuators* **86** 149-153 (2000).

# MICRO-MACHINING OF PZT-BASED MEMS

Todd B. Myers, Susmita Bose and Amit Bandyopadhyay
School of Mechanical and Materials Engineering
Pullman, WA 99164-2920

John D. Fraser
Scanhead Technology Unit
ATL Ultrasound
PO Box 3003, Bothel, WA 98041-3003

## ABSTRACT

PZT thin films were prepared on platinized silicon wafers using sol-gel technique. The films were micro-machined to create structures for MEMS devices. Standard photolithography techniques were used to create structures in silicon and silicon substrate with thin film PZT. This paper will present the process compatibility issues that were encountered during the development of PZT actuated silicon membranes and silicon/silicon dioxide cantilevers. The resulting micro-machined membranes were tested for ferroelectric properties and the performances are compared with properties of thin PZT films on platinized silicon substrates.

## INTRODUCTION

Lead zirconate titanate (PZT) based thin film materials have become a common material of choice for many micro-electromechanical systems (MEMS) because of their excellent piezoelectric properties. When PZT thin film is used in conjunction with a micro-machined substrate, the piezoelectric response of the film can cause the device to produce a mechanical motion due to an electrical input signal. To maximize the mechanical energy output, the influence of all variables including materials and structural designs must be characterized for their electrical as well as mechanical contributions.

Almost all MEMS devices are micro-machined to allow them to perform functions such as, actuation, vibration, or rotation [1-3]. In this work, the structure of interest lies with a silicon membrane actuated by a thin film of PZT. To create membranes with a constant and predictable thickness on a (100) silicon wafer, a boron etch stop was introduced by diffusing boron at a high concentration into one side of the wafer (front-side) [4]. Common anisotropic wet etchants of (100) silicon have etch rates that drop off 50 times upon encountering the heavily doped boron silicon surface ($> 7 \times 10^{19}$ cm$^{-3}$) and still maintain (100)/(111) etch selectivity on the range of 35:1 [5]. The SiO$_2$ is grown on top of the silicon to isolate the membrane from the substrate and act as a structural component. Titanium is used as an adhesion layer for platinum, which is used at a bottom electrode. Platinum offers chemical inertness, as Pt does not oxidize easily, and facilitates columnar grain growth of sol-gel derived PZT. The PZT is deposited using a metallorganic precursor solution because of high quality of the resulting film, low cost, and stoichiometric control.

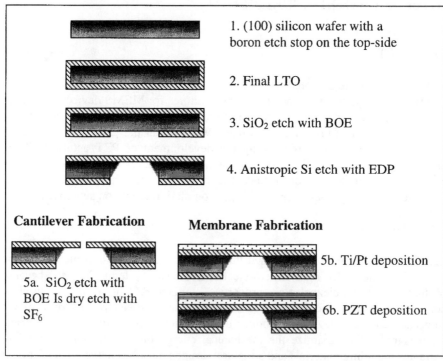

Figure 1: Flowchart showing the microfabrication steps for Si/SiO$_2$ cantilevers and Si/SiO$_2$/Ti-Pt/PZT membranes

Cantilever beam fabrication is also investigated to determine the ability to make cantilevers consisting of the same materials as the membranes. The successful

Innovative Processing and Synthesis of Ceramics

fabrication of heterostructured cantilever beams would allow for the evaluation of mechanical properties of each layer to be fully characterized with some form of mechanical testing, perhaps nano-indentation. Micro-cantilever beams would also allow for the measurement of the transverse piezoelectric coefficient $d_{31}$ [6,7]. Figure 1 shows a schematic of the cantilever and membrane fabrication steps.

Processing Compatibility Issues

In order to fabricate a device with many different materials on the same substrate, as is common in MEMS, the compatibility of all processes with the materials must be well understood. Boron doping and oxidation must be performed first because of the high temperatures used. Anisotropic silicon etching was selected as the second major process to be performed because of the difficulty in maintaining high quality Ti-Pt or PZT during the wet etching of Si using ethylenediamine/pyrocatecol/water (EDP) or tetramethyl ammonium hydroxide (TMAH) or potassium hydroxide (KOH). Wet etching of silicon could have been performed with either TMAH or EDP [8,9] because both the solutions have a high selectivity to oxide masks, are p++ etch stop compatible, and etch at a relatively fast rate. The KOH's oxide selectivity was too high to be utilized when silicon oxide is being used as the mask and also as an important structural component. The TMAH's etch rate is comparable to EDP and is a much benign chemical, though difficulty in obtaining high quality surface finish due to the formation of hillocks on the surface and also TMAH's availability in the lab became major drawbacks. The EDP was selected as the etchant for Si micro-machining because of excellent surface finish, faster etch-rate and availability of the set-up in the lab.

The Ti-Pt deposition must be done in the same deposition chamber to prevent the oxidation of the titanium before the deposition of platinum. Sputtering and electron beam evaporation are both compatible processes. E-beam evaporation was selected because of the option to buy platinum by weight instead of investing in a very expensive platinum target for a sputtering system. Since the PZT deposition is done by spin-coating, the surface being used as a substrate must be very flat and smooth in order to achieve high quality film. The Ti/Pt layer must be deposited upon a smooth flat surface and etched after PZT deposition if the Ti/Pt layer needs to be micro-machined. This approach creates a major obstacle due to difficulty in etching platinum using common dry or wet etching methods. To create the Si/SiO$_2$ cantilevers, a dry etch can be used to cut out a cantilever pattern over a membrane. SF$_6$ can be used as the gas because it is the only available process gas that etches silicon and is compatible with MEMS based processing involving PZT [10].

## EXPERMINTAL PROCEDURE

### Membrane Fabrication

3 inch silicon wafers (p-type <100>, 350-400 um thick) were oxidized at 1050°C in a steam (wet oxidation), oxygen, and nitrogen environment for 2 hours to grow a 500 nm thick oxide layer on Si. The oxide was removed from one side of the wafer using semiconductor tape and a buffered oxide etch consisting of hydrofluoric acid and ammonium fluoride (BOE 10:1). A boron soak was then carried out using boron planer disks at a temperature of 1125°C for 2.5 hours. An oxidation treatment (900°C for 10 minutes) preceded the boron diffusion to assist in the removal of the boron skin that forms on the surface of the silicon. Deglazing was done using BOE 10:1 and a subsequent sacrificial low temperature oxide (LTO) was grown at 850°C for 2 hours in a steam and oxygen environment. The oxide was then removed using BOE 10:1 and a final LTO was grown at 850°C in a steam and oxygen environment for 3 hours. The sacrificial oxide is etched away to improve the surface of the silicon.

Standard photolithography techniques and a contact aligner were used to create an oxide mask on the backside of the wafer consisting of 2.57 mm squares using BOE 10:1 as an etchant. The wafers were then etched with the silicon etchant EDP at 110°C for 5.5 hours resulting in silicon membranes. Upon removal from the EDP, the organic etchant nanostrip was used to ensure a clean surface, free of any organic residue. Wafers designated for membrane fabrication were coated with 20 nm of titanium for adhesion and 200 nm of platinum (bottom electrode) through electron beam evaporation.

### PZT Thin Film

Thin film PZT (52/48) was deposited on to the platinized silicon micro-machined substrate via spin-coating a metallorganic precursor solution. A 0.5 M precursor solution was made using stoichiometric amounts of anhydrous lead acetate, titanium iso-propoxide, and zirconium n-propoxide, with 10% excess lead added to compensate for any lead loss during sintering. Anhydrous lead acetate is produced by the distillation of lead acetate trihydrate and 2-methoxyethonal (2-MOE), which was used as the solvent. A reflux at 120°C for 2-hours was performed after each of the remaining components was added to 2MOE-andyrous lead acetate solution. A final distillation was done to bring the molarity to 0.5 M. Spin-coating is used at 3000 rpm for 12 seconds followed by heat treatments of 150°C for 5 minutes and 350°C for 5 minutes. The 4 layers are applied before a

Innovative Processing and Synthesis of Ceramics

final sintering at 700°C for 10 minutes. Top gold electrodes of 1 mm in diameter were sputtered on the PZT film directly over the membranes and on the bulk substrate using a mask.

## Si/SiO$_2$ Cantilevers

For cantilever beam fabrication, Ti/Pt was not deposited to investigate the influence of the oxide and silicon to the membrane. Photolithography techniques were used to create a cantilever beam pattern on the top-side (boron-doped side) of the wafer, directly over the membranes created from backside etching. The photoresist was then used as a mask for oxide etching in BOE 10:1 and silicon etching using reactive ion etching (RIE, Trion model Phantom I), which was carried out using SF$_6$ gas at 125 W for 2 minutes. The remaining photoresist was removed and a nanostrip etch was performed again to completely clean the surface.

## RESULTS AND DISCUSSION

### Micro-machined Substrate

Silicon oxide thickness were measured with a Nanospec (Nanometrics model 210) on the backside of wafers prior to and during silicon etching to establish etch rates. The etch-rate of the oxide mask during the EDP silicon etching was measured to be 6-8 nm/hr. The etch-rate of the (100) silicon by EDP at 110°C was measured by taking a drop gauge measurement (Heidenhain model CP25M) into partially etched membranes. The measured etch rate was 60-65 µm/hr. Both of these values agree with previously published etch rates for EDP [8]. The final thickness of the micro-machined membranes into the silicon substrate were measured to be 2.75 µm in thickness using measurements taken from the drop gauge and SEM micrographs (JEOL model JSM 6400). SEM micrographs also revealed a very smooth etch surface at the bottom of the membranes indicating a uniform boron doping. When aligned properly to the (110) flats of the (100) wafer, the anisotropic EDP etch revealed the (111) planes as the flat sidewalls of the membranes. **Figures 2a and b** show SEM micrographs of the micro-machined membranes.

### Si/SiO$_2$ Cantilevers

The cantilevers micro-machined from the topside of the membranes were all 2.75 µm in thickness, dictated by the boron diffusion depth. The lengths and widths of the cantilevers fabricated varied from 60-400 µm and 30-60 µm, respectively. A few cantilevers showed a small amount of deflection based upon the appearance

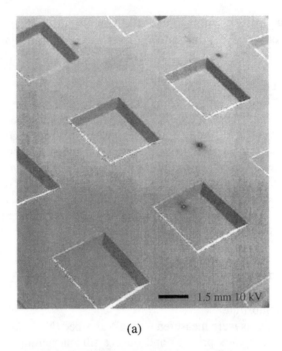

(a)

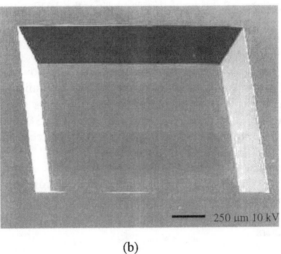

(b)

Figure 2: SEM micrograph of (a) the micro-machined silicon membranes and (b) one micro-machined silicon membrane showing the (111) side-walls and flat membrane surfaces.

from SEM micrographs and subsequent measured lengths. Removal of this oxide through wet etching with BOE tended to compensate some for the deflection, but certainly did not alleviate it completely. These deflections could be caused by residual stresses introduced into the cantilevers during the final oxidation treatment pullout. A furnace cool or a short-term anneal upon the completion of processing would be a better method to minimize these stresses. Another possible cause for the deflection observed, is the cantilevers are not thick enough to support there own weight and therefore deflect. Cantilevers of shorter length (< 250 μm) tended to show no deflection. **Figure 3a and b** showed Si/SiO$_2$ micromachined cantilevers.

(a)

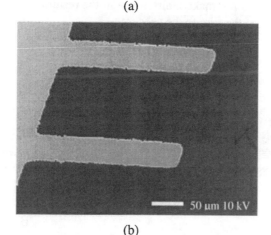

(b)

Figure 3: (a) Micro-machined cantilevers measuring 300 and 400 μm in length, 45 μm in width, and 2.75 μm thick. There is a 200 nm LTO on the surface. (b) Two 300 μm cantilevers at a higher magnification.

Ferroelectric Properties of PZT on Micro-Machined Substrates

A profilometer (Tencor model Alpha Step 100) measured the thickness of the 4-layer PZT thin film to be 280 nm. Ferroelectric testing (Radiant Technologies model RT66A) was done on PZT thin films deposited on platinized silicon substrates to compare hysteresis loops of thin film PZT on a Si/SiO$_2$/Ti-Pt/PZT/Au substrate (Si thickness +350 μm) and on Si/SiO$_2$/Ti-Pt/PZT/Au membrane (Si thickness 2.75 μm). Two different wire-bonding methods were used to make contact with the top electrode. The needle probe placed directly onto the top electrode did not provide good electrical contacts due to the displacement of the membrane as measurements were taken.

In the first method, graphite paste was placed onto the top electrode between a needle probe and the top electrode on the film. This method produces a semi-clamped electrode for the film because some stress from the needle probe is transferred to the film. The graphite paste acted as a displacement buffer, allowing for a constant electrical contact to be made during testing. The second method involved bonding a copper wire (diameter 40 μm) with graphite paste to the top electrode and making the electrical contact through the wire. This method provided the PZT thin film with an unclamped electrode, transferring no stress because of contact to the top electrode was through a very thin wire that did not inflict any mechanical stress. **Figure 4** shows a schematic of the two different contact methods used, A and B. The contact method for thin film on a non-machined substrate did not make a difference in the resulting P-E hysteresis loop due to the clamping of the film to the substrate.

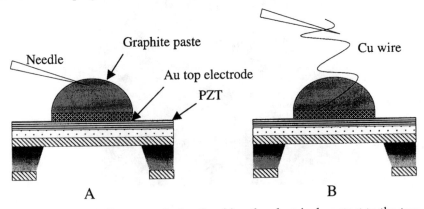

Figure 4: Two different methods of making the electrical contact to the top electrode of the PZT thin film. A is semi-clamped top electrode and B is an unclamped top electrode.

Innovative Processing and Synthesis of Ceramics

The semi-clamped top electrode PZT on a membrane and PZT on a silicon substrate gave similar ferroelectric properties. The hysteresis loops for the two cases are shown in **Figure 5a**. The coercive field ($E_c$), remnant polarization ($P_{rem}$), and saturation polarization ($P_s$) all show the similar values. This implies that a film over a membrane can still constrained similar to a film on a substrate by the application of a rigid top electrode.

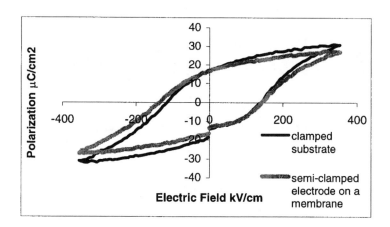

Figure 5a: Hysteresis loops for clamped thin film and a semi-clamped electrode on top of a 2.75 μm membrane.

**Figure 5b** shows a comparison of hysteresis loops, in which both measurements are recorded on a membrane. The $P_{sat}$ and $P_r$ values of the unclamped electrode showed considerably higher values as opposed to the semi-clamped electrode. This may be due to the fact that the unclamped electrode transfers no stress to the PZT-membrane and allows it to displace completely. The $E_c$ value remained same for the two methods tested.

CONCLUSION

Sol-gel derived PZT thin films were prepared on platinized silicon wafers. Standard photolithography techniques were used for micro-machining to create structures in silicon and silicon substrates with thin PZT film for MEMS devices. This paper discussed issues related to the process compatibility that were encountered during the development of PZT actuated silicon membranes and silicon/silicon dioxide cantilevers. The resulting micro-machined membranes were tested for ferroelectric properties. It was found that measurement techniques play a significant role on the properties. The PZT thin film properties when measured under semi-clamped condition, the substrate thickness did not play any significant role. However, the same properties showed extensive variations when measured unclamped. The Si/SiO$_2$ cantilevers were fabricated for future mechanical property measurement using devices such as nano-indenters. Some residual stress related bending was observed on as processed high aspect ratio cantilevers.

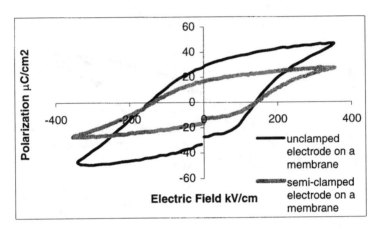

Figure 5b: Semi-clamped electrode on a membrane vs. an unclamped membrane

ACKNOWLEDGEMENTS

The authors would like to acknowledge the financial support from the Washington Technology Center, Seattle and ATL Ultrasound, Bothel, WA. Experimental support from Mr. Oliver Huang and Mr. Gabe Kuhn is highly appreciated. Useful discussion with Mr. Parag Banerjee, Dr. David Bahr, and Mr. Kevin Bruce is also acknowledged.

REFERENCES

1. W.P. Robbins "Ferroelectric-Based Microactuators" *Integrated Ferroelectrics*, **11** ¼ 179-191 (1995)
2. J. Golebiowski "Fabrication of Piezoelectric thin film of Zinc Oxide in composite membrane of ultrasonic microsensors" *Journal of Material Science*. **34** [19] 4661-4664 (1999)

3.  Y. Suzuki, K. Tani, T. Sakuhara "Development of a New Type Piezoelectric Motor" *Sensors and Actuators*. **83** 244-288 (2000)
4.  E. Manea, R. Muller, A. Popescu "Some Particular Aspects of the Thin Membrane by Boron Diffusion Processes" *Sensors and Actuators* **74** 91-94 (1999)
5.  G.T.A. Kovacs "Wet Etching of Silicon" pp. 31-47 in *Micromachined Transducers Sourcebook*. McGraw-Hill, New York 1998
6.  M. Toyama, R. Kubo, E. Takata, K. Tanaka, K. Ohwada "Characterization of Piezoelectric Properties of PZT Thin Films Deposited on Si by ECR Sputtering" *Sensors and Actuators A*. **45** 125-129 (1994)
7.  P. Luginbuhl, G. Racine, P. Lerch, B. Romanowicz, K. Brooks, N. de Rooij, P. Renaud, N. Setter "Piezoelectric Cantilever Beams Actuated by PZT Sol-Gel Thin Film" *Sensors and Actuators A*. **54** 530-535 (1996)
8.  H. Seidel, L. Csepregi, A. Heuberger, H. Baumgartel "Anisotropic Etching of Crystalline Silicon in Alkaline Solutions" *J. Electrochem Soc.* **137** [11] 3612-3626 (1990)
9.  U. Schnakenberg, W. Benecke, P. Lange "TMAHW Etchants for Silicon Micromachining" *1991 International Conference of Solid State Sensors and Actuators* IEEE June 24-28 (1991) 815-818
10. S.P. Beeby, A. Blackburn, N.M. White "Processing of PZT Piezoelectric Thick Films on Silicon of Microelectromechanical Systems" *J. Micromech. Microeng.* **9** 218-229 (1999)

Innovative Processing and Synthesis of Ceramics

# TEMPLATED GRAIN GROWTH OF LEAD METANIOBATE CERAMICS BY SOLID FREEFORM FABRICATION

Kazuhiro Nonaka[*], Mehdi Allahverdi, and Ahmad Safari
Department of Ceramic and Materials Engineering
Center for Ceramic Research
Rutgers, The State University of New Jersey
607 Taylor Road
Piscataway, NJ 08854-8065

## ABSTRACT

The feasibility of fused deposition of ceramics (FDC) process, a unique solid freeform fabrication (SFF) technique to fabricate grain oriented ceramic components using templated grain growth technique was investigated in lead metaniobate ($PbNb_2O_6$; PN) ceramic. Needle-shaped PN seeds (2-5 $\mu$m-diameter, 10-50 $\mu$m-length) and equi-axed PN powder (0.3 $\mu$m-average diameter) were prepared by the molten salt and conventional solid-state reaction processes, respectively. Fine equi-axed powder and 5 wt% seeds were mixed and compounded with a thermoplastic binder to prepare suitable feedstock for FDC with a solids loading of 52 vol%. The compounded materials were then extruded into filaments of 1.78 mm-diameter. SEM analyses showed that the seeds survived harsh compounding during filament preparation without major damage. The seeds were also aligned in the extrusion direction upon filament fabrication. In the components deposited from the filaments using FDC, the seeds retained their alignment parallel to the deposition direction. The FDC components showed a highly oriented microstructure with ferroelectric orthorhombic phase upon sintering at 1300 °C for 1h. XRD pattern collected from the surface of such samples showed that all the diffraction peaks from (hkl) planes with l≠0 were missing in the pattern, implying the strong grain orientation in a plane parallel to the deposition layer.

## INTRODUCTION

Templated grain growth (TGG) technique has been extensively studied for

[*]Permanent address: National Institute of Advanced Industrial Science and Technology (AIST), Kyushu 807-1 Shuku-machi, Tosu, Saga 841-0052 Japan   (E-mail: k.nonaka@aist.go.jp)

various ceramic systems including ferroelectrics such as $Bi_4Ti_3O_{12}$ (BiT)[1] and its family[2], as well as structural ceramics such as alumina[3], mullite[4], and $Si_3N_4$[5]. The objective was to obtain improved properties by development of textured microstructure. In the TGG process, single crystal-like particles with needle or platelet shape (i.e. seeds) are used as templates for anisotropic grain growth of fine matrix particles. In this process, the primary alignment of seeds in a desired direction plays a key role in the development of strong orientation in the final processing stage. As the most common technique for alignment of seeds, tape casting method has been utilized by many researchers[1-5]. Anisotropic seed particles, dispersed in a slurry of fine powder and a low-viscosity medium, can be oriented during casting. The alignment occurs due to shear stresses applied on the slurry, forcing the seeds to re-orient in the casting direction. Upon sintering and TGG processes, grain oriented microstructures can be developed throughout the tapes.

Fused deposition of ceramics (FDC)[6,7], a solid freeform fabrication process, allows rapid prototyping of 3-D ceramic components directly from CAD files in a short time and low cost. In this process, feedstock materials (powder-loaded polymer filaments) are deposited through a fine nozzle, along a computer-controlled tool-path. During FDC, deposited materials are exposed to moderate shear stresses, which can be used to obtain primary seed alignment. Therefore, if anisotropic particles, such as needles and platelets, are included in the filaments, they can be oriented in the deposition direction. Prior to this work, FDC experiments combined with TGG process were carried out using BiT particles with platelets morphology. It was observed that the platelet particles were aligned on the surface of the filaments parallel to the extrusion direction.[8] However, in the bulk of BiT filaments, the platelets were loosely oriented. In the FDC components built with seeded BiT filaments, the orientation of the platelets in a plane parallel to a deposition layer was improved owing to further shearing of the material during deposition. Upon sintering and TGG processes, a moderately oriented microstructure with Lotgering factor of 45 % was developed. This experiment manifested the potential of FDC technique for development of textured microstructures.

It is evident that the orientation of the seeds during extrusion depends strongly upon the morphology of the seeds. For example, if needle-shaped particles are used in the FDC filaments, a proper alignment is expected both on the surface and within the bulk of the filaments. In this paper, we report on the feasibility of the FDC-TGG processes to prototype net shape grain oriented ceramic components of lead metaniobate (PN) ceramics. It is known that needle-shaped PN particles can be synthesized via molten salt technique.[9] Using such seeds, one directional, grain oriented microstructure was fabricated by 2-stage hot pressing.[10] The resulted samples showed improved piezoelectric properties along pressing directions. Similar results were also reported upon tape casting process.[11]

As the most promising application of PN ceramics, electro-acoustic applications (hydrophones) should be noted, where a higher $d_{33}/d_{31}$ ratio and hence a higher $k_t/k_p$ ratio is beneficial.[12] In polycrystalline samples, high $d_{33}/d_{31}$ ratio can be obtained when the microstructure shows strong grain orientation (as opposed to a random granular structure). Hence, if PN samples can be fabricated with grain-orientated microstructures, improved electro-acoustic properties are expected.

## EXPERIMENTAL

### Powder Preparation

*Seeds*: Needle-shaped particles of PN were synthesized as templates by molten salt method. Equimolar mixtures of PbO and $Nb_2O_5$ powders were mixed with an equal weight of KCl, and ball milled for 24 h using zirconia media and isopropyl alcohol. The mixtures were heated at 1150 °C for 1h in alumina crucibles to grow PN seeds. The KCl was washed out of the fused KCl-seeds by decantation with hot deionized water more than 15 times. The washed samples were filtered and dried at 110 °C for 24 h.

*Matrix Powder*: Fine, equiaxed particles of PN were prepared by conventional solid-state reaction process as matrix powders. Equimolar PbO and $Nb_2O_5$ powders were ball milled for 24 h with similar media used for seeds preparation. The mixture was dried, calcined at 800 °C for 1 h, and ball-milled for 48 h. The resulting fine powder was filtered and dried at 110 °C for 24 h.

### FDC Process

*Compounding and Filament Extrusion*: Prior to making filaments, the matrix powder was coated with stearic acid as a surfactant to be able to uniformly disperse the powder in a thermoplastic binder (ECG-9).[13] The coating process was performed by adding the powder to a toluene solution containing 3 wt% stearic acid (4.5 wt% vs. matrix powder) and mixing for 24h.[14]

The coated powder was compounded in a high shear mixer (Haake Record 9000) with the ECG-9 binder (polyolefin base) at 140-160 °C to obtain 52 vol% PN solids loading. To minimize damage during compounding process, the seeds were added into the mixer at a later stage without coating treatment.

PN filaments of 1.78 mm-diameter were extruded from the compounded materials using a Rheometer attached to Instron 4500 at 81 °C.

*FDC Components*: In the FDC process, the material is extruded through a fine nozzle onto a fixtureless platform. The deposition head (liquefier) moves in the a horizontal (x-y) plane according to computer-controlled tool-path. To prototype PN components, the filaments were fed into the liquefier heated at 145-160 °C. The layer by layer (254 µm-thickness/layer) deposition was carried out until a complete green component was built.

**Sintering**

The green PN components were subjected to a slow binder burnout cycle up to 550 °C at a heating rate of 10 °C/h with duration of 2 h at the maximum temperature. The resulting samples were embedded in PN powder, placed in an alumina crucible, sealed with alumina cement, and then sintered under a PbO atmosphere at 1000-1300 °C for 1h (heating rate of 6 °C/min), followed by furnace cooling.

**Characterization**

The crystalline phases formed were identified by X-ray diffraction (XRD; Siemens D500) using CuKα radiation with Ni filter. Densities of sintered samples were measured by Archimedes method using water. The microstructures of the samples were observed with a scanning electron microscope (SEM; Amray Corp., Model 1200).

**RESULTS AND DISCUSSION**

**Fabrication of Filaments and FDC Components**

Figure 1 shows the SEM micrographs of PN powders synthesized as matrix and seeds. The matrix powder has equiaxed particles with 0.3 μm in average diameter, while the seeds consist of needle-shaped particles with 2-5 μm-diameter and 10-50 μm-length. The major aspect ratio of the seeds is estimated to be about 10. The XRD measurements showed that the crystal structure of the seeds is orthorhombic whereas the matrix powder revealed a rhombohedral structure. It has been reported that in the needle-shaped PN particles prepared by molten salt synthesis, the long axis of the needles is oriented parallel to the c-axis (or [001] axis) of PN crystal.[10,11]

**Fig. 1.** SEM micrographs of PN (a) matrix powder and (b) needle-shaped seeds.

**Fig. 2.** SEM micrographs of seeded (5 wt%) PN filaments upon binder burnout at 550 °C for 1h: (a,b) filament surfaces and (c,d) cross sections of the filament at low and high magnifications.

SEM micrographs taken from the FDC filaments are shown in Figure 2. As shown, the seeds can be seen both on the surface and fracture surface of the filament with proper alignment along extrusion direction. It should be noted that during compounding and extrusion, seeds are normally exposed to high shear stresses, which can be damaging. However, it was found that the seeds maintained their original morphology without a heavy damage in the extruded filaments. This is important since the survival of seeds is a critical factor in the success of the following TGG step to produce grain-oriented microstructure.

Figure 3 shows the SEM micrographs of FDC extrudate (i.e. fine stream of material extruded through FDC nozzle). As previously seen in Fig. 2 (b), the seeds are aligned in the extrusion direction, similar to the FDC filaments. This observation demonstrates that alignment of the seed can be maintained and transferred to FDC component.

SEM micrographs of the FDC components are illustrated in Figure 4. The seed particles both on the surface and within the bulk are mainly oriented parallel to the deposition direction. This indicates that the near net-shape components with primary seed alignment can be prototyped by the FDC process.

**Fig. 3.** Low- and high-magnification SEM micrographs of fine steams of seeded PN material extruded through FDC nozzle. The samples were heat treated (550 °C for 1h) to remove the binder.

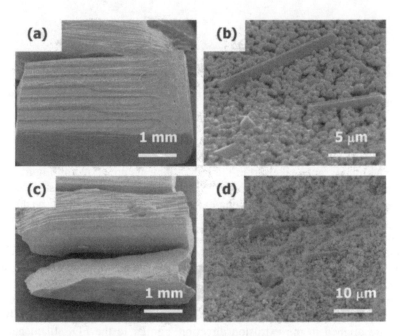

**Fig. 4.** SEM micrographs of the FDC components after binder burnout: (a,b) surfaces, and (c,d) fractured surfaces at low and high magnifications.

**Microstructure Development**

Figure 5 shows the microstructure development of the seeded FDC samples and those of unseeded disks prepared by uni-axial pressing as a comparison. In

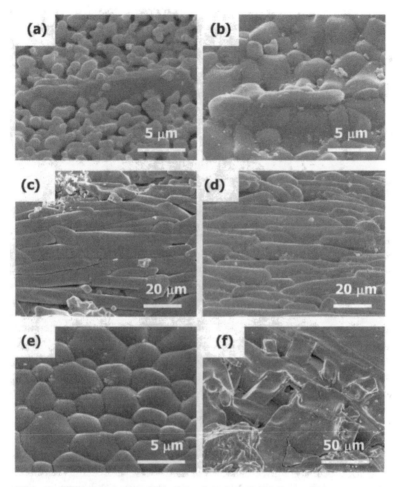

**Fig. 5.** SEM micrographs of (a-d) seeded FDC components and (e,f) uni-axially pressed disks sintered at 1100 °C (a), 1200 °C (b,e), 1250 °C (c) and 1300 °C (d, f) for 1h.

the samples sintered at low temperatures, the seeds initially started to sinter to the surrounding matrix grains (Fig. 5 (a)). Densification reaction proceeds rapidly in a narrow temperature range between 1100 °C and 1200 °C, as plotted in Figure 6. The microstructure developed at 1200 °C consisted of matrix grains of mostly equi-axed shape (Fig. 5 (b)). When the sintering temperature increased to 1250 °C, a remarkable change in the microstructure was observed. All of the grains had grown into elongated grains with large aspect ratios, and mainly

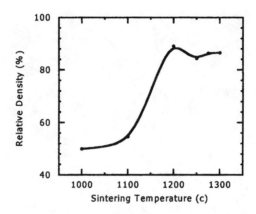

**Fig. 6.** Relative density of the seeded FDC components vs. sintering temperature.

oriented parallel to the deposition direction (Figs. 5(c) and 5(d)). On the other hand, the microstructures of the uni-axially pressed disks consisted of uniform and randomly grown grains upon sintering at 1200 °C and 1300 °C, respectively. These results suggest that templated grain growth of PN using needle-shaped seeds can be well achieved via the FDC process.

As seen in Fig. 5 (c), some large cracks and pores are also observed in the samples. These imperfections are probably caused by the extensive growth of elongated grains (>50 μm-length). A significant grain growth may lead to large and localized thermal stresses between grains due to anisotropic thermal expansion during cooling. Also, misalignment of the elongated grains may result in grain boundaries separation with further grain growth, resulting in pore/crack formation. Accordingly, the densification curve shows a maximum at 1200 °C, followed by a decrease in density with increasing temperature to 1250 °C. Comparing the microstructures of the two FDC samples sintered at 1250 °C and 1300 °C (Figs. 5 (c) and 5(d)), no significant difference is seen in grain size and aspect ratio.

**XRD Phase Analysis and Grain Orientation**

XRD patterns collected from the sintered FDC samples are plotted in Figure 7 as a function of sintering temperature. All the data were measured on as-sintered surfaces with the exception of pattern (d), which was obtained from a ground sample for a comparison. To retain the ferroelectric phase upon cooling to room temperature, the samples must be sintered at temperatures higher than 1250 °C.[12] A single ferroelectric phase with orthorhombic structure was obtained by sintering at 1300 °C. Only the diffractions peaks of (hk0) planes appear in the as-sintered sample. The peaks corresponding to (hkl) planes with

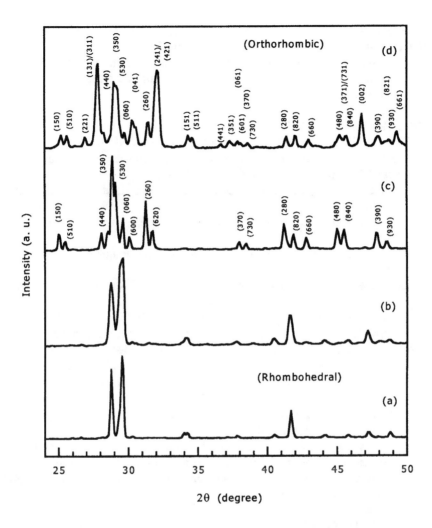

**Fig. 7.** XRD patterns of the seeded FDC components sintered at 1200 °C (a), 1250 °C (b) and 1300 °C (c ,d) for 1h. All the patterns were collected from as-sintered surfaces with the exception of (d), which is obtained from the powdered sample.

$l \neq 0$ were completely disappeared in the pattern. On the other hand, all the peaks reported in the Powder Diffraction File (PDF) card[15] were identified for the ground sample. These results indicate that the needle-like grains are largely orientated in a plane parallel to the deposition layer, and are in good agreement with the result of microstructural observation (Fig. 5 (d)).

To obtain finer microstructure with similar grain orientation, we are currently studying sintering process in more detail and evaluate the effect of seed content on the microstructure.

## SUMMARY AND CONCLUSIONS

The feasibility of development of grain-oriented microstructure was investigated in PN ceramic components using FDC and templated grain growth techniques. We summarize our findings as follows:

1. Fine and equiaxed PN particles were synthesized via oxide mixing method.
2. Needle-shaped seeds were prepared by molten salt synthesis using KCl flux.
3. The needles survived in PN filaments without major damage and were aligned in the extrusion direction.
4. In green FDC components, the seeds were also oriented parallel to depositing direction.
5. During sintering process at temperatures higher than 1250 °C, the matrix particles grow rapidly into elongated grains with high aspect ratios.
6. Upon sintering at 1300 °C for 1h, grain-oriented microstructure with ferroelectric orthorhombic phase was obtained in the FDC components.
7. XRD pattern collected from the surface of the samples sintered at 1300 °C showed that all (hkl) diffraction peaks with $l \neq 0$ were missing, indicating strong grain orientation of c-axis direction in a plane parallel to the FDC deposition layer.
8. The FDC process can successfully be employed to prototype components with grain oriented microstructure. This is done by primary alignment of needle-shaped seeds in the green samples, followed by templated grain growth at elevated temperatures.

## ACKNOWLEDGEMENT

The authors wish to acknowledge ONR for funding of the project (N00014-96-1-0959). We would also like to thank Yoni Berkowitz, undergraduate student of Rutgers University, for his assistance in experimental procedures.

## REFERENCES

1. J.A. Horn, S.C. Zhang, U. Selvaraj, G.L. Messing and S. Trolier-McKinstry, "Templated Grain Growth of Textured Bismuth Titanate", *Journal of the American Ceramic Society*, **82** [4] 921-26 (1999).

2. T. Takeuchi, T. Tani and Y. Saito, "Piezoelectric Properties of Bismuth Layer-Structured Ferroelectric Ceramics with a Preferred Orientation Processed by the Reactive Templated Grain Growth Method", *Japanese Journal of Applied Physics*, **9B** [38] 5553-56 (1999).

3. E. Suvaci, M.M. Seabaugh and G.L. Messing, "Reaction-based Processing of Textured Alumina by Templated Grain Growth", *Journal of the European Ceramic Society*, **19** 2465-74 (1999).

4. S.H. Hong and G.L. Messing, "Development of Textured Mullite by Templated Grain Growth", *Journal of the American Ceramic Society*, **82** [4] 867-72 (1999).

5. K. Hirao, M. Ohashi, M. Brito and S. Kanzaki, , Processing Strategy for Producing Highly Anisotropic Silicon Nitride", *Journal of the American Ceramic Society*, **78** [6] 1687-90 (1995).

6. A. Bandyopadhyay, R. K. Panda, V.F. Janas, M.K. Agarwala, R. van Weeren, S.C. Danforth and A. Safari, "Processing of Piezocomposites by Fused Deposition Technique"; pp.999-1002 in *Proceedings of the Tenth IEEE International Symposium on Applications of Ferroelectrics (ISAF) '96*, Part vol.2, Edited by B.M. Kulwicki, A. Amin and A. Safari. IEEE, East Brunswick, 1996.

7. S.C. Danforth, M.K. Agarwala, A. Bandyopadhyay, N. Langrana, V.R. Jamalabad, A. Safari and R. van Weeren, "Solid Freeform Fabrication Methods", U.S. Pat. No. 5 738 817, Apr. 1998.

8. M. Allahverdi, B. Jadidian, Y. Ito and A. Safari, "Fabrication of Bismuth Titanate Components with Oriented Microstructures via FDC and TGG", in *Proceedings of the twelfth IEEE International Symposium on Applications of Ferroelectrics (ISAF) 2000* (in press).

9. T. Kimura, T. Yamaguchi and R.E. Newnham, "Phase and Morphology of $PbNb_2O_6$ Obtained by Molten Salt Synthesis", *Particulate Science and Technology*, **1** 357-64 (1983).

10. K. Nagata and K. Okazaki, "One-Directional Grain-Oriented Lead Metaniobate Ceramics", *Japanese Journal of Applied Physics*, **24** [7] 812-14 (1985).

11. M. Granahan, M. Holmes, W.A. Schulze and R.E. Newnham, "Grain-Oriented $PbNb_2O_6$ Ceramics", *Journal of the American Ceramic Society*, **64** C-68-9 (1981).

12. P. Eyraud, L. Eyraud, P. Gonnard, D. Noterman and M. Troccaz, "Electromechanical Properties of $PbNb_2O_6$ and $PbTiO_3$ Modified Ceramics Elaborated by a Coprecipitation Process"; pp.410-13 in *Proceedings of the Sixth IEEE International Symposium on Applications of Ferroelectrics*, IEEE, New York, 1986.

13. T.F. McNulty, F. Mohammadi, A. Bandyopadghyay, D.F. Shanefield, S.C. Danforth and A. Safari, "Development of a Binder Formulation for Fused

Deposition of Ceramics", *Rapid Prototyping Journal*, **4** [4] 144-50 (1998).

14. T.F. McNulty, D.J. Shanefield, S.C. Danforth and A. Safari , "Dispersion of Lead Zirconate Titanate for Fused Deposition of Ceramics", *Journal of the American Ceramic Society*, **82** [7] 1757–60 (1999).

15. Powder Diffraction File (PDF) No. 70-1388.

# INVESTIGATION OF NI-CU-ZN FERRITE WITH HIGH PERFORMANCE DERIVED FROM NANO FERRITE POWDERS

Xiaohui Wang, Weiguo Qu, Longtu Li, Zhilun Gui, Ji Zhou
State Key Lab of New Ceramic & Fine Processing
Department of Materials Science & Engineering
Tsinghua University
Beijing, 100084, P. R. China

ABSTRACT

$(Ni_{0.2}Cu_{0.2}Zn_{0.6}O)_{1.01}(Fe_2O_3)_{0.99}$ nanocrystalline powders were synthesized by a citrate precursor method. The nanocrystals with grain sizes in the range from 10nm to 60nm were obtained at various calcining temperatures. Because of excellent sintering activity, the nanocrystals can be sintered below 900°C without the addition of sintering aids. The sintered body possesses fine-grained microstructure, showing high initial permeability up to 700 with high DC resistivity of $10^9 - 10^{10}$ ohm · cm. Influences of initial particle size as well as sintering temperature on microstructures and properties of sintered body were studied. The results show that the nanocrystalline Ni-Cu-Zn ferrite is a promising material for high frequency MLCI application.

INTRODUCTION

In the past ten years, with the development of surface mounting technology, great progresses have been made in the miniaturization of electromagnetic components. As one of the most important media materials, Ni-Cu-Zn ferrite has been commercially used to manufacture multilayer chip inductor in high frequency region. In order to be cofired with the inner electrode (usually Ag is the most suitable internal contact material with a melting point of 961°C), Ni-Cu-Zn ferrite must be sintered below 900°C. The normal way to lower the sintering temperature is using sintering aids such as $Bi_2O_3$, $V_2O_5$ etc, which usually does great harm to the magnetic property of the ceramics acquired [1-4]. Ultrafine Ni-Cu-Zn ferrite with uniform composition, small grain size and narrow size distribution is expected to realize low-temperature sintering and attain ideal magnetic property at the same time [5].

Many methods have been developed to prepare ultrafine powder including co-precipitation, hydro-thermal method, peptization method, sol-gel method, crystal glass method, metallorganics hydrolyse method, microlatex method and so on [6-10]. In this paper, Ni-Cu-Zn ferrite nanocrystals with a composition of $(Ni_{0.2}Cu_{0.2}Zn_{0.6}O)_{1.01}(Fe_2O_3)_{0.99}$ were prepared via a citrate precursor method. The Ni-Cu-Zn ferrite nanocrystals with different grain sizes were obtained by controlling the heat-treatment conditions. The crystalline structures, sintering ability and magnetic properties of these powders were investigated.

EXPERIMENTAL PROCEDURE
Sample Preparation
The starting materials were iron nitrate, nickel acetate, copper acetate and citric acid. Firstly, iron nitrate was dissolved in distilled water before being precipitated by a ammonia solution to form $Fe(OH)_3$. Filtered, washed, the fresh $Fe(OH)_3$ precipitate was dissolved into hot citric acid solution at 60~80°C in the ratio of 1:1 of Fe: citric acid, and a transparent solution was obtained. Then, nickel and copper acetate were added in stoichiometric quantities in the above solution to give the required composition. After that appropriate ammonia was drop into the above sol until it was neutral or slightly alkaline (pH=6-8). After heating at 125°C, dried gel were obtained. Finally, the gel was ignited or heat treated at 600°C, 700°C and 800°C for 4h, respectively , resulting Ni-Cu-Zn nanocrystals with different grain sizes .

The nanocrystalline powders were pre-milled and pressed into disks (10mm in diameter and 1.0mm thickness) and toroidal samples (20mm outside diameter, 10mm inside diameter and 3mm thickness) with the addition of 5wt% PVA as lubricant. The ceramic samples sintered at 870~890°C were used for magnetic property measurements.

Characterization and Property Measurements

The crystalline structures were determined by a Rigaku Powder X-Ray diffractometer using Fe K$_a$ radiation. The morphology and grain size of the nanocrystalline powders were observed using a JEM-200CX transmission electron microscopy (TEM). The magnetic properties of the ceramics were measured using an HP 4194A Low Impedance Analyzer. The DC resistivity of the ceramic was determined using an HP 4040B Micro-amperemeter. The microstructures of the ceramics were investigated using a Zeiss CSM-950 scanning electron microscopy (SEM).

RESULTS AND DISCUSSION
Phase and Structure Analysis

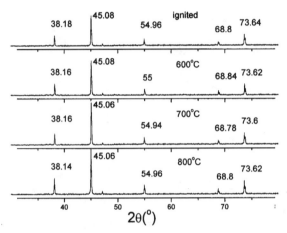

Figure 1.    XRD patterns of nanocrystalline powders
ignited or    heat-treated at different temperatures

Figure 1 shows XRD patterns of nanocrystalline powders ignited or calcined at different temperatures. It is obvious that a single spinel phase was formed for all the powders, and the intensity of XRD peaks increased with increasing calcination temperature indicating that the crystals had been developed with the increase of heat-treatment temperature.

Grain Size and Morphology
    The average grain sizes of the nonocrystalline Ni-Cu-Zn powders were calculated according to the full width at half maximum (FWHM) of the strongest peak [11]. Table I  gives the average grain size data for all the powders.

Table I. The average grain sizes of the various powders

| Sample No. | 1 | 2 | 3 | 4 |
|---|---|---|---|---|
| Heat-treatment condition | ignited | 600℃/4h | 700℃/4h | 800℃/4h |
| Average grain size (nm) | 11(XRD) 8 (TEM) | 18 (XRD) 20 (TEM) | 26 (XRD) 30 (TEM) | 60 (XRD) 80 (TEM) |

Figure 2 shows the morphology of the nanocrystalline powders. The particles of all the powders were spherical in shape with uniform grain size. The grain sizes of the powders determined by TEM are also listed in table 1. With increasing heat-treatment temperature, the grains grew up.

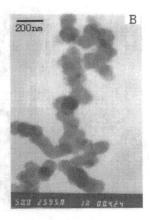

Figure 2     TEM photographs of ferrite nanocrystals
(A) Ignited powder (150 000×)
(B) Powder calcined at 800℃ (50 000×)

Magnetic and Electrical Properties

Due to the high surface area, the Ni-Cu-Zn nanocrystals prepared by this method show obviously excellent sintering activity. It could be sintered below 900 ℃ without the addition of sintering aids, avoiding the relevant deterioration of properties caused by sintering aids. The ceramics produced from nanocrystalline powders displayed better magnetic and electrical properties in comparison with those made by conventional ceramics method.

*Permeability-frequency characteristics:* figure 3 shows frequency-dependent permeability curves of ceramic samples sintered at various temperatures. These ceramic samples are all derived from the powders calcined at 700 ℃. For comparison, a curve of ceramic prepared by conventional ceramics method is also presented in this figure. It is obvious that the initial permeability of the samples increased with the increasing sintering temperature. This could be explained by the effect of sintering temperature on the grain size. As the sintering temperature increases, the grains of the samples tend to grow and thus the domain rotation becomes much easier. The most important result is that the initial permeability of the sample sintered at 890℃ was about 700, which is really hard to be achieved using conventional ceramic method.

Figure 4 gives the permeability-frequency curves for the ceramics sintered at 890℃ derived from the powders calcined at different temperature (with various grain sizes). Obviously, the initial permeability does not change linearly with the increasing heat-treatment temperature. On the contrary, it increases first and then drops.

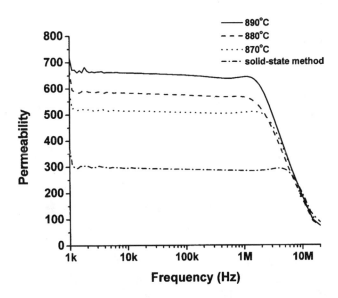

Figure 3.    Frequency-dependent permeability curves of samples sintered at different temperatures

The most preferable result can be acquired by the sample derived from the powder calcined at 700℃ with average grain size of 30nm. This behavior can also be explained by the size effect of Ni-Cu-Zn nanocrystals. At the beginning, an increase in grain size of the nanocrystals can help to enlarge the ceramic grain size and reduce the resistance of domain rotation, thus increasing the initial permeability. But as the grain size of the nanocrystals increases further, the specific surface area and the surface activity decrease, which may lead to lower density of the sintered ceramic. That may be the reason for the drop of the initial permeability of the sample made from powders calcined at 800℃ with larger grain size (as shown in figure 4).

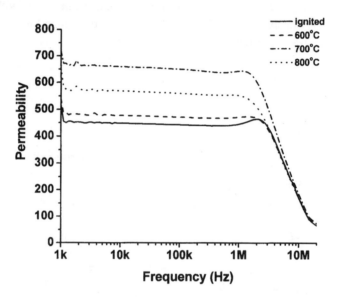

Figure 4.    Frequency-dependent permeability curves of ceramics sintered at 890℃ derived from powders calcined at different temperatures

*DC resistivity property:* an adequate DC resistivity of the ferrite is also required in manufacturing of MLCI so as to avoid "creeping" while electroplating, which may do great harm to the quality of components. Usually, DC resistivity of $10^7 \sim 10^8$ ohm • cm can be achieved by solid-state method which can hardly fit the requirement. Whereas, ceramics sintered from the nanocrystals prepared by this method show high DC resistivity of $10^9 \sim 10^{10}$ ohm • cm, which is much more preferable for practical production and for the reliability of the components. Figure 5 shows the SEM photos of the ceramic samples derived from the nanocrystals and that from usual powder prepared by solid-state method. It is obvious that the grains of the ceramic derived from nanocrystals are small, uniform and compacted with each other. As well known , the most important factor that affects the resistivity of ceramic is the proportion of grain boundaries in it. Obviously, this proportion is much higher for the ceramic derived from nanocrystals than that made by conventional method. That is why the ceramic sintered from the nanocrystals has much higher resistivity.

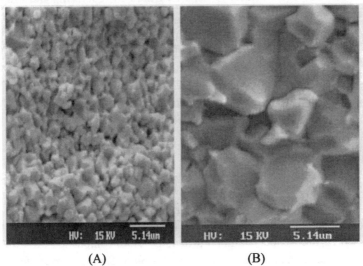

(A)                              (B)

Figure 5. SEM micrographs of ceramics derived from:
(A)    nanocrystalline powder (5000×)
(B)    powder prepared via a solid-state reaction (2000×)

CONCLUSIONS

Ni-Cu-Zn ferrite nanocrystals were successfully prepared via a citrate precursor method. Low temperature (below 900℃) sintering of Ni-Cu-Zn ferrite is realized without doping any sintering aids using the nanocrystalline powders. The magnetic and electrical properties of the ceramics derived from the nanocrystals are much better than those prepared by a conventional ceramic method. Therefore, the nanocrystalline Ni-Cu-Zn ferrite is a promising material for high frequency MLCI application.

ACKNOWLEDGEMENT

This work was    supported by the National Natural Science Foundation of P. R. China under grant No.59995523.

REFERENCES

[1] Yamaguchi T, Shinagawa M, "Effect of Glass Addition and Quenching on the Relation Between Inductance and External Compressive Stress in Ni-Cu-Zn Ferrite Glass Composites", *Journal of Materials Science*, **30** [2] 504-508 ( 1995).

[2] Jean JH, Lee CH, "Effects of Lead(II) Oxide on Processing and Properties of Low-Temperature-Cofirable", *Journal of the American Ceramic Society*, **82** [2] 343-350 (1999).

[3] Hsu JY, KO WS, Chen CJ, "The Effect of $V_2O_5$ on the Sintering of NiCuZn Ferrite", *IEEE Transactions on Magnetics*, **31** [6] 3994-3996 (1995).

[4] He XH, Xiong MR, Ling ZY, Qiu QC, "Low-temperature Sintering of NiCuZn Ferrite for Multilayer-Chip Inductor", *Journal of Inorganic Materials*, **14** [1] 71-77 ( 1999).

[5] Date SK, Deshpande CE, "Magnetism in Materials - Processing and Microstructure", *Metals Materials and Processes*, 7 [1] 15-18 ( 1995).

[6] Yue ZX, Zhou J, Li LT, Zhang HG, Gui ZL, "Synthesis of Nanocrystalline NiCuZn Ferrite Powders by Sol-Gel Auto-combustion Method", *Journal of Magnetism and Magnetic Materials*, **208** [1-2] 55-60 (2000).

[7] Yue ZX, Li LT, Zhou J, Zhang HG, Gui ZL, "Preparation and Characterization of NiCuZn Ferrite Nanocrystalline Powders By Auto-combustion of Nitrate-citrate Gels", *Materials Science and Engineering B - Solid State Materials for Advanced Technology*, **64** [1] 68-72 (1999).

[8] Kim WC, Park SI, Kim SJ, Lee SW, Kim CS, "Magnetic and Structural Properties of Ultrafine Ni-Zn-Cu Ferrite Grown by a Sol-Gel Method", *Journal of Applied Physics*, **87** [9] 6241-6243, Part 3 ( 2000).

[9] J.F.Hochepied, "Nonstoichiometric Zinc Ferrite Nanocrystals: Syntheses and Unusual Magnetic Properties", *J.Phys.Chem.B* **104**   905~912 (2000).

[10] Hsiao-Miin Sung, Chi-Jen Chen, "Fine Powder Ferrite for Multilayer Chip Inductors", *IEEE Transactions on Magnetics*, **30** [6] 4906~4908 ( 1994).

[11] C. Suryanarayana and M. Grant Norton, *"X-Ray Diffraction – A Practical Approach"*; pp.207-, 1998

# KEYWORD AND AUTHOR INDEX